ABC's of Hydraulic Circuits

by

Harry L. Stewart

and

John M. Storer

HOWARD W. SAMS & CO., INC.
THE BOBBS-MERRILL CO., INC.
INDIANAPOLIS • KANSAS CITY • NEW YORK

FIRST EDITION

FIRST PRINTING—1973

International Standard Book Number: 0-672-21003-7
Library of Congress Catalog Card Number: 73-83372

Preface

The purpose of this book is to familiarize the reader with basic hydraulic circuit building techniques, beginning with a fundamental explanation of symbols and symbology—the basic tools and universal language of fluid power. The control circuits in this book range from the simple to the complex, from single-motion control to multiple interlocks, and from individual manual actuations to automatic sequencing and programming. Multiple approaches to specific problems point toward simplicity of design, facility of maintenance, and dependability of performance. Many specific circuits designed to perform particular tasks are depicted. By concentrating on explanation and orientation to specific applications, the end product is a wide variety of performance-tested circuits that are ready to be put to work meeting the everyday needs of industry.

HARRY L. STEWART
JOHN M. STORER

Contents

CHAPTER 5

CHAPTER 6

CHAPTER 7

CHAPTER 8

CHAPTER 9

CHAPTER 10

CHAPTER 11

CHAPTER 1

Tools of Hydraulic Circuit Building

Anyone who wishes to design, develop, or troubleshoot hydraulic circuits can do so by following certain basic fundamentals. Among these are:

1. Understand the significance and limitations of the basic building blocks of hydraulic circuits as employed in standard practice by circuit designers.
2. Keep the hydraulic system design as simple as possible by not overcrowding it with extra controls that do not serve a particular purpose.
3. Realize that hydraulic circuits employ a relatively incompressible fluid at high pressures, making it conducive to such conditions as heart, shock, leakage, restrictions, etc. This can be a source of major problems if proper circuit design is not employed.
4. Consider the type of fluid to be employed when the circuit is to be placed in operation; this may result in a change in some of the basic components of the system.
5. Consider both the operating temperature and ambient temperature in relation to the system; they could have a bearing on the functioning of the system.
6. Consider location of reservoirs, heat exchangers and heaters, drain lines, and exhaust lines.

PICTORIAL SYMBOLS

Historically, three types of symbols have been used to design fluid power circuits. Pictorial symbols may be as simple as a square, rectangle, or circle used to represent a component. Usually, however, they take on the configuration of the envelope of the component and are made to represent the external appearance of the component (Fig. 1-1). They are comparatively easy to draw and are satisfactory when used merely to indicate the plumbing lines and the perspective location of components. Pictorial symbols are rather inadequate for conveying the functional aspects of the components within a circuit.

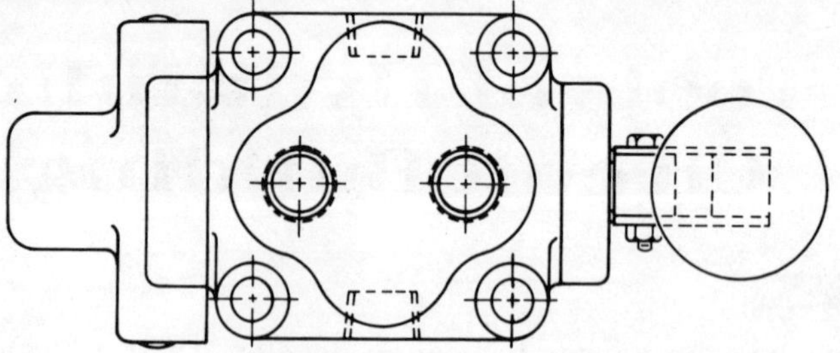

Fig. 1-1. Example of an outline illustration.

CUTAWAY SYMBOLS

Cutaway symbols have also been used for circuit drawing. These are quite difficult to draw and mainly serve the purpose of depicting the construction of the component (Fig. 1-2). The circuit designer often finds the majority of his efforts being spent on properly depicting the construction of the component, thus detracting from his efforts to register and create the functional aspect of the circuit. Furthermore, cutaway symbols do not take into account the variety of constructions that may be encountered in the industry. For the purpose of this study, neither the pictorial-type symbol nor the cutaway type will be used.

ANS GRAPHIC SYMBOLS

All symbols employed hereafter are patterned after those established as ANS Graphic Symbols for fluid power diagrams. These graphic symbols can depict, in very simple easy-to-draw form, the functional details of a component as well as the

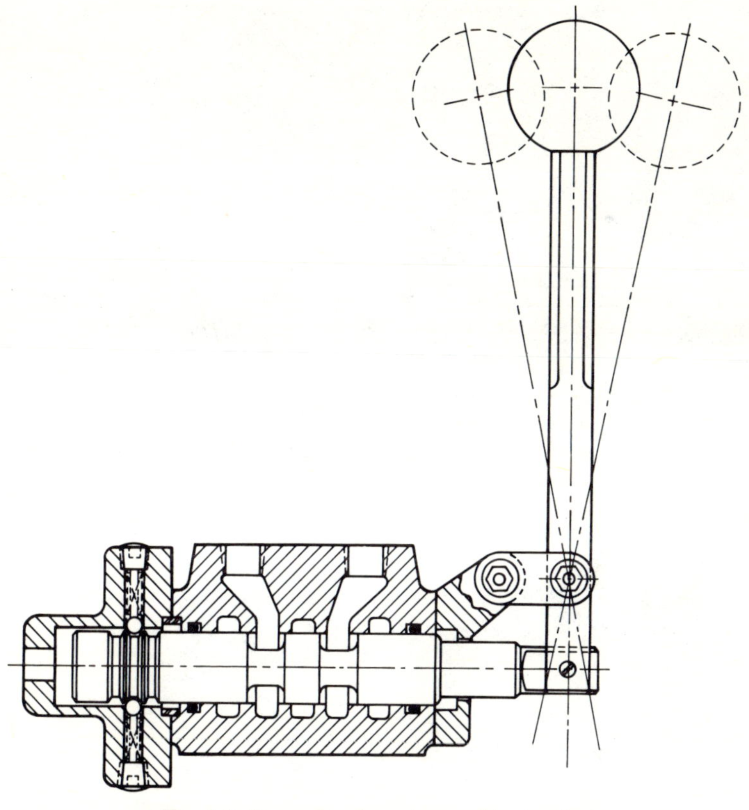

Fig. 1-2. Example of a cutaway illustration.

method of operation of that component. Careful detail has been given to the design of these graphic symbols, allowing simple representation of all details that are essential to the understanding and creation of a hydraulic circuit. While these symbols are often used in more than one form, simplified forms are most popular and are the easiest to work with. An effort will be made to point out some, although not all, of the differences between the simplified symbol and its more complex form. In some cases, the more complex form is more practically employed in depicting the actual function of a component within a hydraulic circuit. In the case of pilot-operated solenoid valves, however, the simplified symbols are easier to employ and, in most cases, are quite adequate.

These simplified symbols are made up of easily reproduced figures consisting of circles, squares, triangles, rectangles, arcs, arrows, lines, dots, and crosses. While some lettering and de-

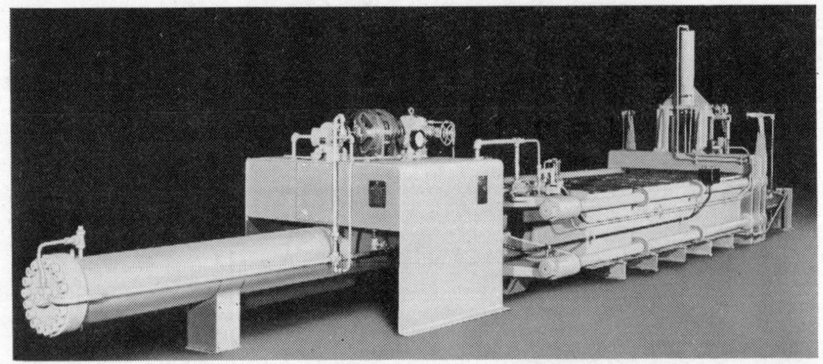

Courtesy HPM Division of Koehring Company.

Fig. 1-3. Large rayon steeping press equipped with hydraulically actuated end gate to facilitate unloading the dehydrated pulp. Note the amount of hydraulic equipment.

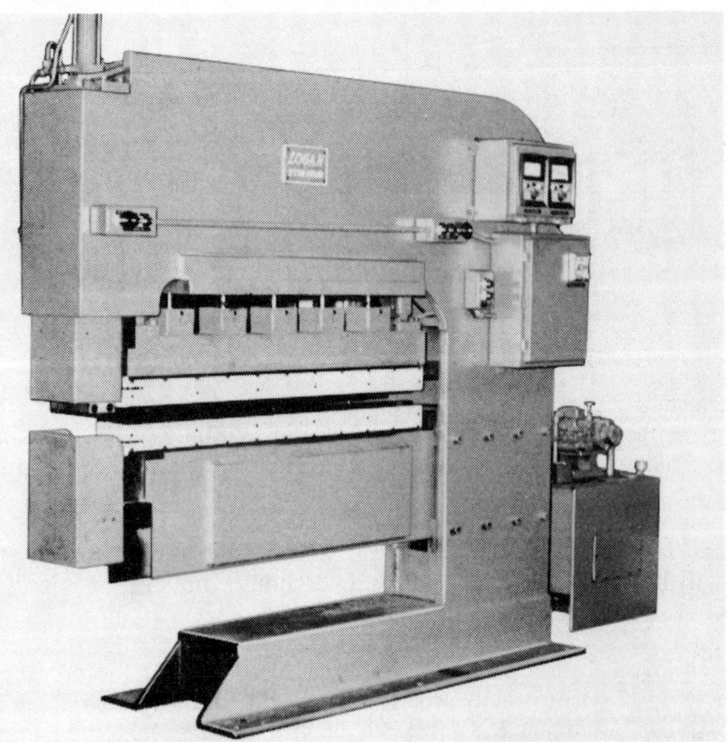

Courtesy of Logensport Machine Co., Inc.

Fig. 1-4. Hydraulic operated sealing press with heated platens. On this press are a group of cylinders that must operate in unison. The cylinder on top is a locking cylinder.

Courtesy International Harvester Company.

Fig. 1-5. An all-purpose standard duty hydraulic loader designed for use with International industrial tractors. Note the hydraulic components used in the system.

Courtesy of The Oilgear Company.

Fig. 1-6. Large hydraulic power unit equipped with pumps, electric motors, and hydraulic controls. Note the size of the oil reservoir and the various access openings.

Courtesy of Industrial Metal Products Corp.

Fig. 1-7. Machine used in the automotive industry which is equipped with a large number of hydraulic power units and large hydraulic cylinders.

scriptive words may be used for clarity, they are unnecessary. The complete omission of words and phrases creates, thereby, a universal language that is able to cross national or cultural barriers. The net result is a universal language of its own, often referred to as "The Universal Language of Fluid Power."

COMBINATIONS

Once the basic figures are mastered, any composite symbol can be devised by using combinations of these figures. Graphic symbols are designed to show connections, flow paths, and functions of components within a circuit. They can indicate conditions occurring during transition from one flow path arrangement to another. They are not intended to indicate construction, nor do they indicate values, such as pressure or flow rate. They are not intended to show locations of ports, the direction of the shifting of spools, or the relative position of actuators on the actual component. Although the symbols may be drawn to any suitable size, the size of the symbol is not intended to indicate difference in component port size.

ANS SYMBOLS

The following are the most widely used ANS Graphic Symbols for hydraulic circuitry:

Lines

The width of the lines does not affect the meaning of the symbols; line widths should be nearly equal for all symbols.

Solid line. A main line conductor, outline, or shaft.

Dash line. A *pilot* line for a control.

Dotted line. A *drain* line.

Center line. An outline for an *enclosure.*

Lines *crossing* (not necessarily at a 90° angle), but *not connected.*

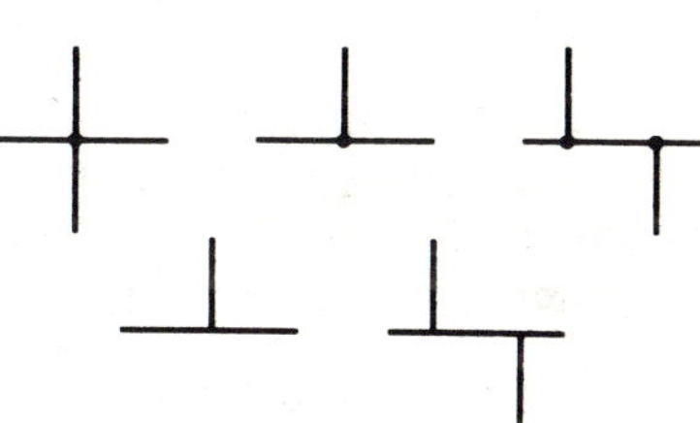

Lines *joining.* Usually indicating a *pressure line* and a *connector.*

Basic Symbols

The basic symbols can be any suitable size. Sizes can be varied for emphasis or clarity; however, relative sizes should be maintained.

Circles and semicircles.

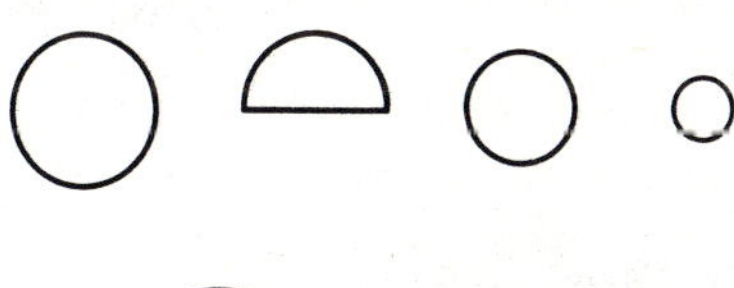

Large and small circles can be used to indicate that a component is the *main* component and the other is an *auxiliary* component.

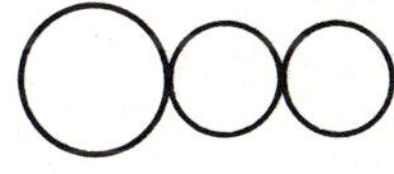

Triangle.

Arrow.

Square and rectangle.

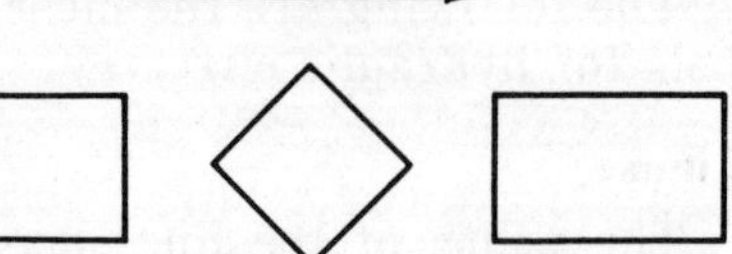

Letter combinations are sometimes used with basic symbols, but they are not necessarily abbreviations. In multiple-envelope symbols, the flow condition nearest an actuator symbol occurs when the control is permitted or caused to actuate. Each symbol shows the normal, at-rest, or neutral condition of the component, unless multiple diagrams are provided to show the various phases of circuit operation.

An arrow at approximately 45° through a symbol indicates that the component can be adjusted or varied.

An arrow parallel to the short side of a symbol, and inside the symbol, indicates a pressure-compensated component.

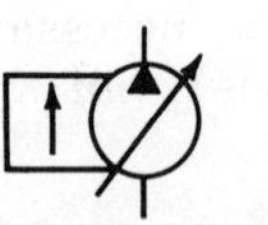

A line terminated in a dot represents a thermometer and is the symbol for temperature cause or effect.

An arrow (on the near side of a shaft) indicates the direction of rotation of a rotating shaft.

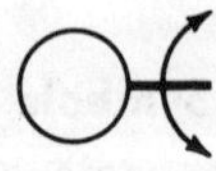

FLUID CONDUCTORS

These lines can be seamless steel, aluminum or copper tubing, or iron pipe.

Pressure Lines

A *working* line (main) or pipe carrying either air or oil under pressure.

Hydraulic line, indicating *direction* of flow.

A *pilot line* (for control), usually running from a pilot valve to a master valve; these lines are usually small in diameter and are made from the same materials used in pressure lines.

A *flexible* line, usually a rubber hose, such as is used to connect a pivot-mounted cylinder.

A line with a *fixed* restriction.

A pressure line with a *plug* connection.

A liquid *drain* line.

Quick-Disconnect

Conductor with a *rotating coupling*.

Conductor *without* checks *connected* (upper) and *disconnected* (lower).

Conductor with *two* checks *connected* (upper) and *disconnected* (lower).

Conductor with *one* check *connected* (upper) and *disconnected* (lower).

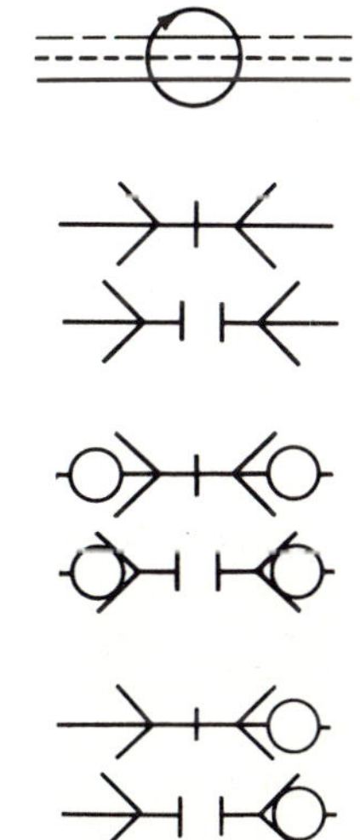

FLUID STORAGE AND ENERGY STORAGE

Conventionally, reservoirs are drawn in the horizontal plane. All lines enter and leave the reservoir from above. A line entering or leaving below the reservoir is shown only when the bottom connection is essential to the circuit function.

Reservoir

A *vented* reservoir or tank.

A *pressurized* reservoir or tank.

Connecting lines enter and leave *above* fluid level.

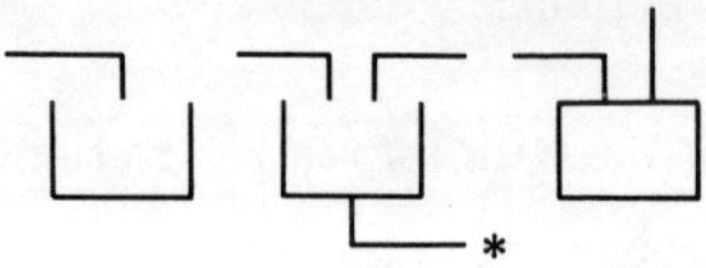

*

Connecting lines enter and leave *below* fluid level.

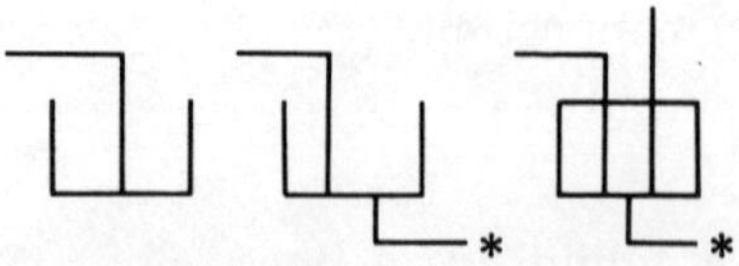

* *

Simplified symbol. Lines enter and leave *above* fluid level. (The return line terminates at the upright legs of the tank symbol.)

Simplified symbol. Lines enter and leave *below* fluid level.

Vented manifold.

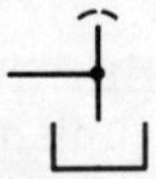

Accumulator.

Accumulator, *spring-loaded.*

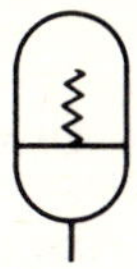

Accumulator, *gas-charged.*

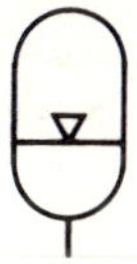

Accumulator, *weighted.*

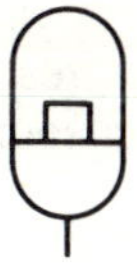

Receiver for air or other gases.

Hydraulic energy sources (pump, accumulator, etc.)

Example →

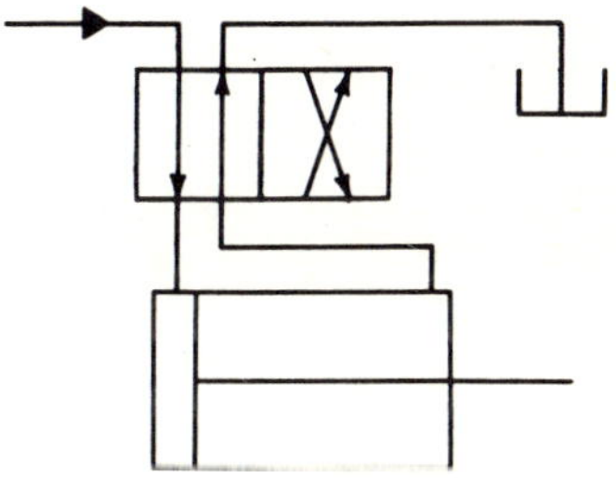

FLUID CONDITIONERS

Device for controlling the physical characteristics of the fluid.

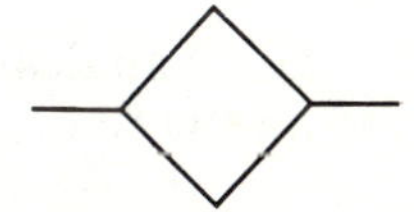

Heater. The triangles indicate that heat is introduced.

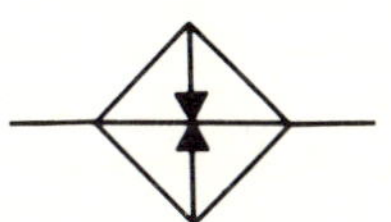

Heater. Outside triangles indicate a *liquid* heating medium.

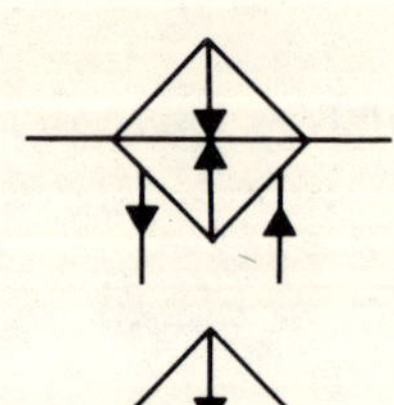

Heater. Outside triangles indicate a *gaseous* heating medium.

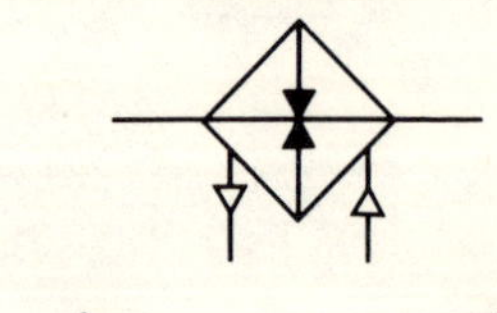

Temperature controller. Outside triangles indicate a liquid or gaseous medium.

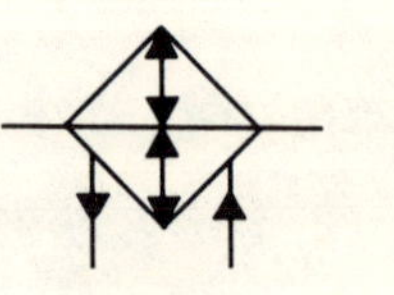
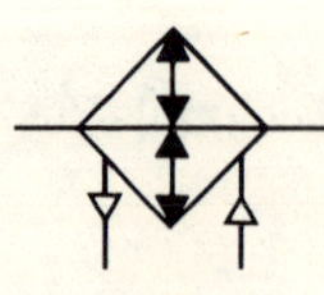

Temperature controller. The temperature is maintained between two predetermined limits.

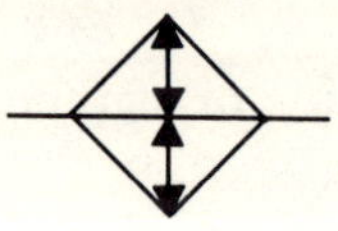
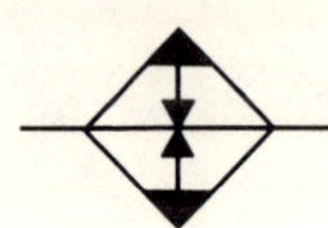

Lubricator *without* drain.

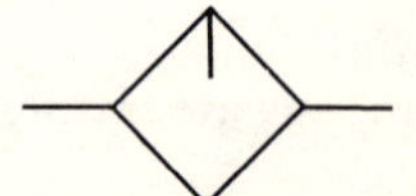

Lubricator with *manual* drain.

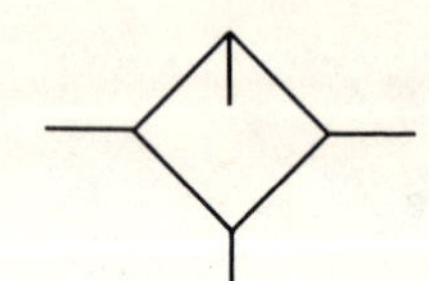

Cooler. Outside triangles indicate a liquid or gaseous medium.

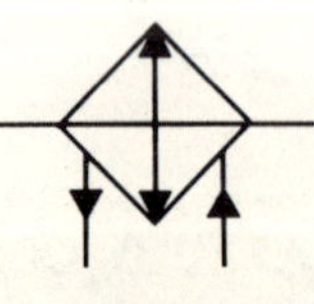
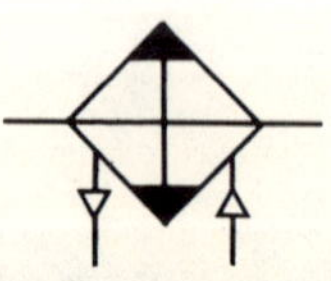

Cooler. Inside triangles indicate heat dissipation.

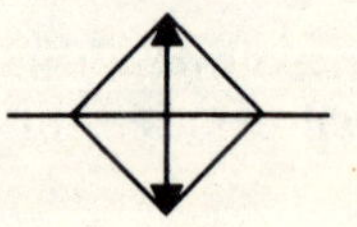
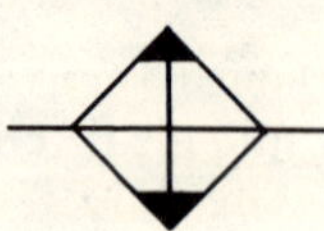

Filter or strainer.

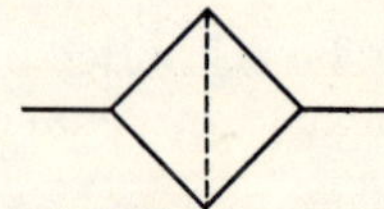

LINEAR DEVICES

Cylinders (Hydraulic and Pneumatic)

Pressure booster or *intensifier.*

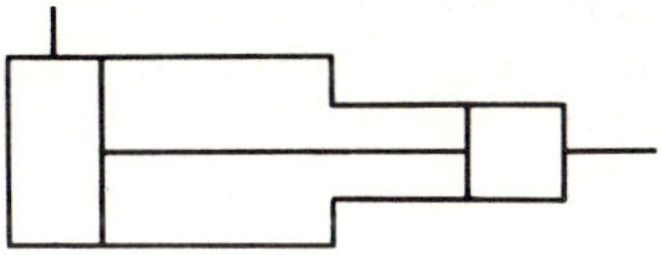

Single-acting cylinder.

Double-acting cylinder with *single end* rod.

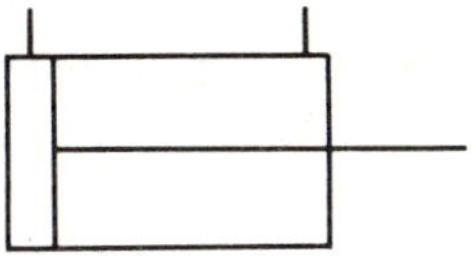

Double-acting cylinder with *double end* rod.

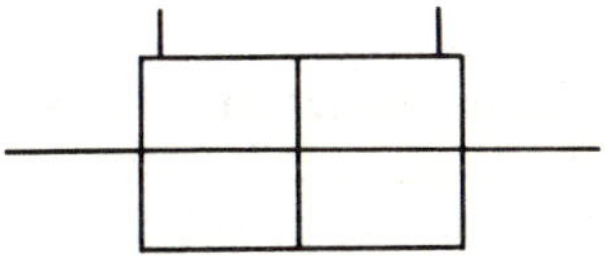

Double-acting cylinder with *fixed* cushion, *advance and retract.*

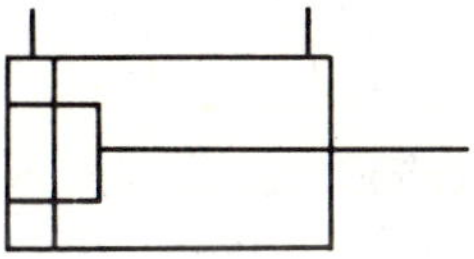

Double-acting cylinder with *adjustable* cushion, *advance only.*

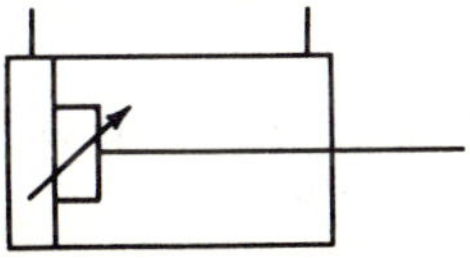

Double-acting cylinder with *fixed cushion, advance and retract* (use this symbol when rod diameter compared to bore diameter is important to function of circuit).

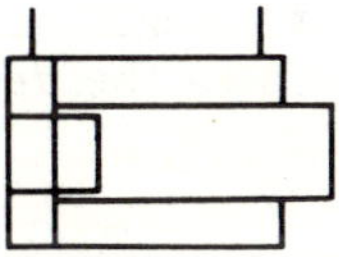

Double-acting cylinder *without cushion* (use this symbol when diameter of rod compared to diameter of bore is important to function of circuit).

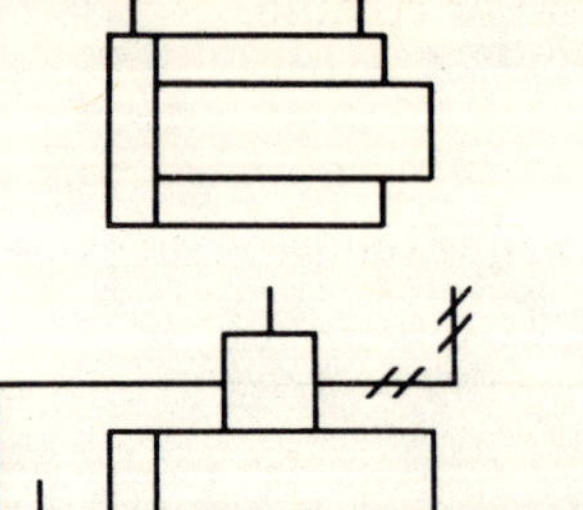

Hydraulic *servo positioner* (simplified).

Discrete positioner (combination of two or more basic symbols).

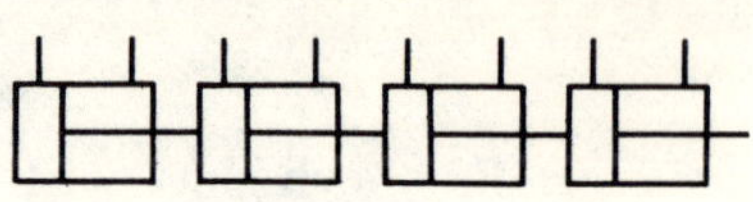

ACTUATORS AND CONTROLS

Manual (A general symbol without indicating a specific type, such as foot, hand, leg, arm, etc.)

Spring actuator or control.

Electrical actuator or *solenoid-actuated* (single-winding) control.

Electrical actuator or *reversing motor.*

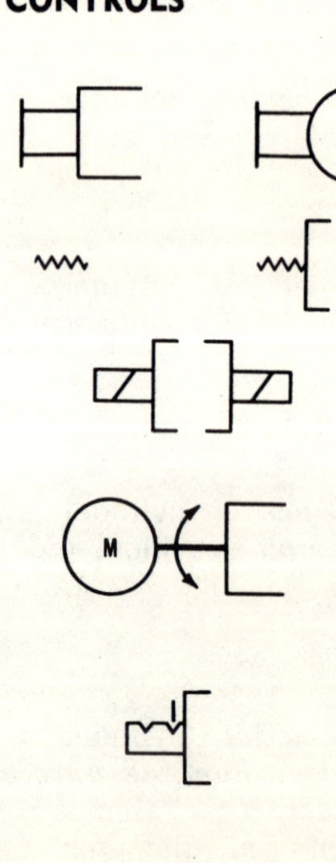

Detent. A notch should indicate each detent in the actual component. A short line indicates the *detent in use.*

Lever.

Pedal or treadle.

Push button.

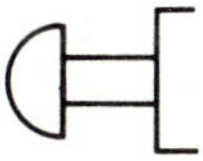

Mechanical.

Pressure-compensated.

Device actuated by *solenoid and pilot* pressure.

Thermal with bulb for *remote sensing*.

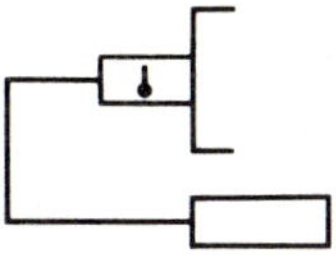

Device actuated by a *servo* (symbol represents energy input, command input, and resultant output).

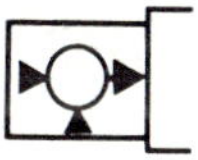

Device actuated by *pilot pressure* (remote supply).

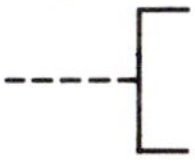

Device actuated by *pilot pressure* (internal supply).

Device actuated by *released pilot pressure* (remote exhaust and internal return).

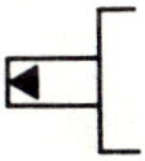

One signal *and* a second signal are required to actuate the device.

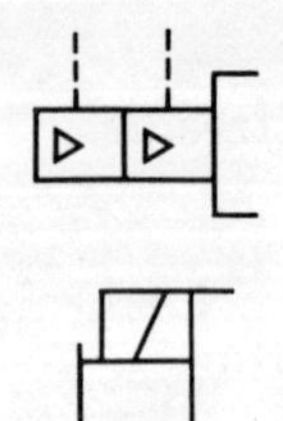

One signal *or* the other signal is required to actuate the device.

ROTARY DEVICES

Basic symbol for a *rotary* device.

Rotary device *with ports.*

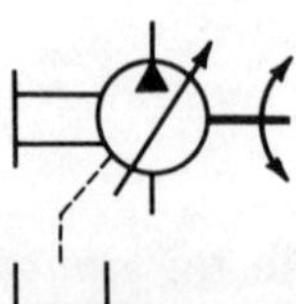

Rotary device with a rotating shaft, a control, and a drain.

Unidirectional fixed-displacement hydraulic pump.

Bidirectional fixed-displacement hydraulic pump.

Unidirectional noncompensated variable-displacement hydraulic pump.

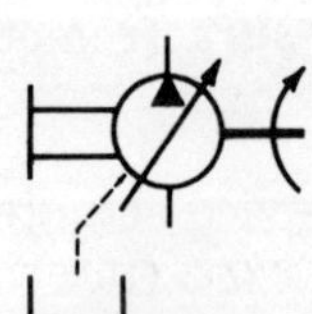

Bidirectional noncompensated variable-displacement hydraulic pump.

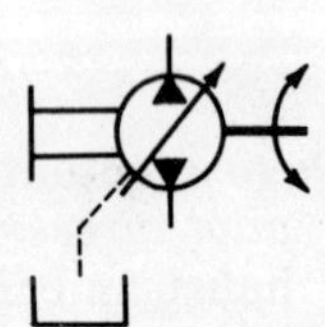

Spring-centered pilot-controlled actuator.

Spring-centered pilot-controlled actuator (simplified symbol).

Device actuated by *pilot-differential* pressure.

Device actuated by pilot-differential pressure (simplified).

Device actuated by *solenoid or pilot pressure* (external pilot supply).

Device actuated by *solenoid or pilot pressure* (internal pilot supply and exhaust).

Thermal. A mechanical device which responds to thermal change (*local sensing*).

One signal *only* is required to actuate the device.

The solenoid *and* the pilot or the manual override *alone* is required to actuate the device.

The solenoid *and* the pilot or the manual override *and* the pilot are required to actuate the device.

The solenoid *and* the pilot or a manual override *and* the pilot or a manual override *alone* can be used to actuate the device.

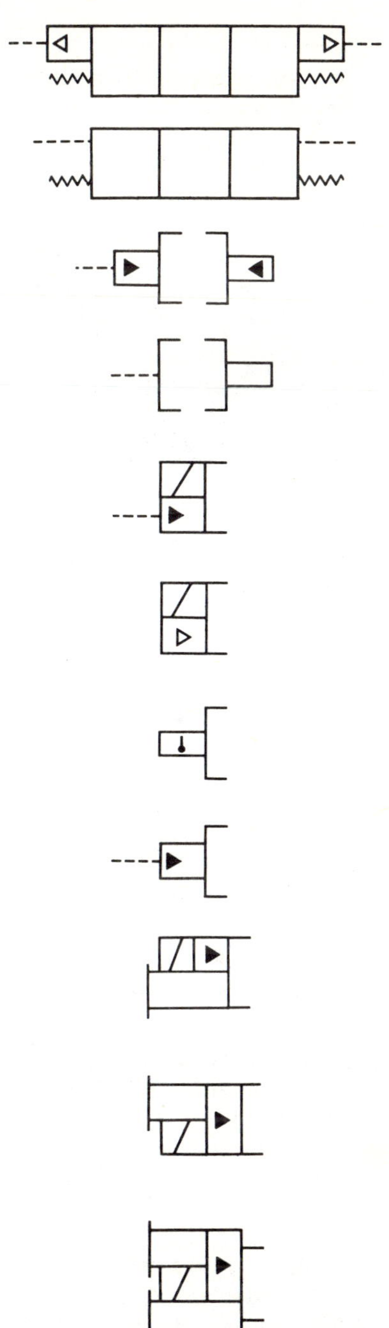

Unidirectional *pressure-compensated* variable-displacement hydraulic pump and simplified symbol.

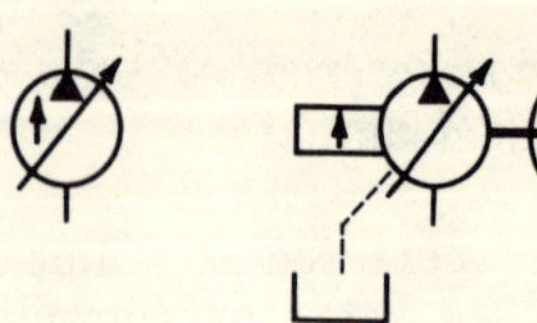

Bidirectional *pressure-compensated* hydraulic pump and simplified symbol.

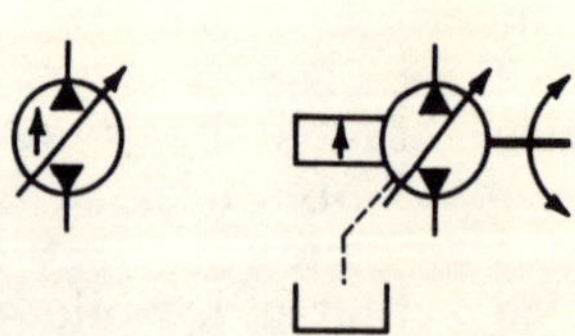

Unidirectional *fixed-displacement* hydraulic motor.

Bidirectional fixed-displacement hydraulic motor.

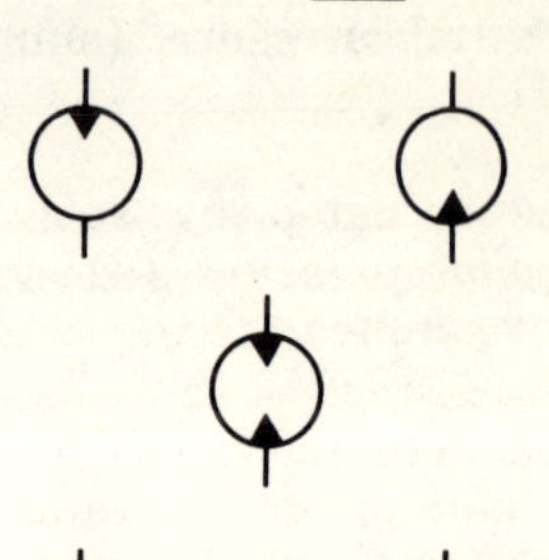

Bidirectional variable-displacement hydraulic motor.

Unidirectional *variable-displacement* hydraulic motor.

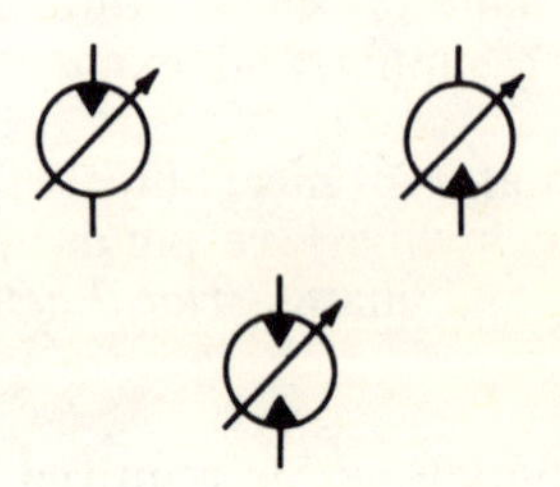

Hydraulic pump-motor, operating in *one* direction as a pump and in the *opposite* direction as a motor.

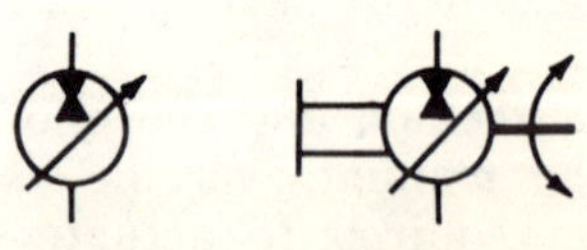

Hydraulic pump-motor, operating in *both* directions of flow either as a pump or as a motor (pressure-compensated variable-displacement).

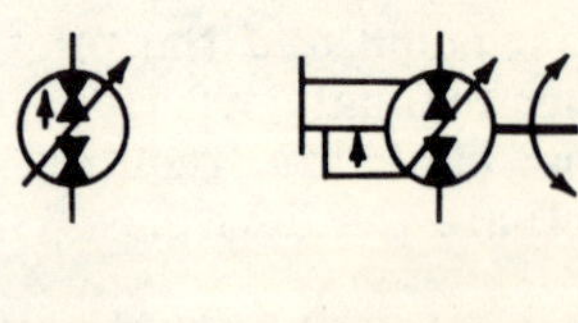

Hydraulic pump-motor operating in *one* direction of flow, either as a pump or as a motor.

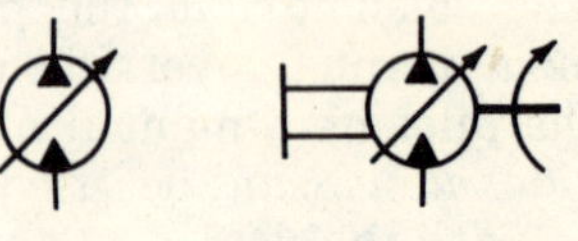

Hydraulic *oscillator*.

Electric motor.

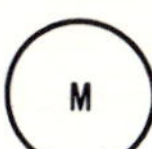

Heat engine (internal combustion, steam).

INSTRUMENTS AND ACCESSORIES

Pressure-indicating and pressure-recording instruments.

Temperature-indicating and temperature-recording instruments.

Flow-rate meter.

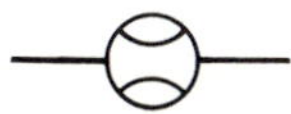

Totalizing meter.

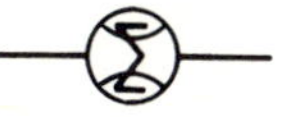

Venturi.

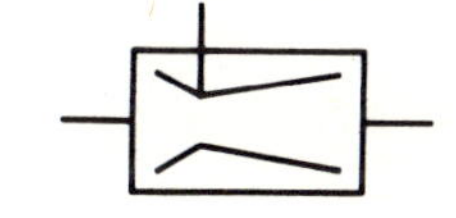

Orifice plate.

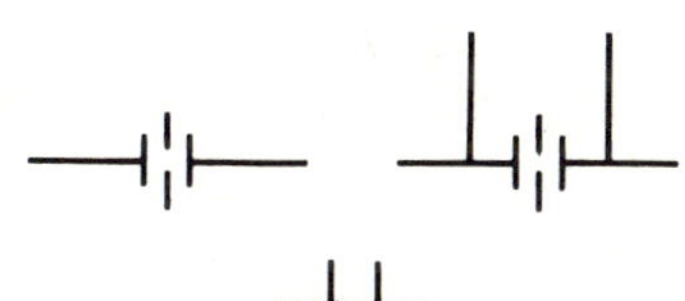

Pitot tube.

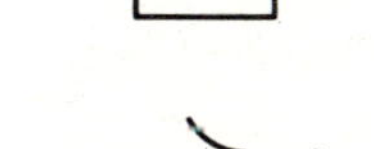

Nozzle (hydraulic).

Pressure switch.

VALVES

A basic valve symbol composed of one or more envelopes with lines inside the envelope to represent flow paths and flow conditions between ports. Three symbol systems are used to represent valve types; (1) single envelope, both finite and infinite positions; (2) multiple envelope, finite position; and (3) multiple envelope, infinite position. The symbol systems are:

1. In finite-position single-envelope valves, the envelope is imagined to move to represent how pressure or flow conditions are controlled as the valve is actuated.
2. Multiple envelopes are used to symbolize valves providing more than one finite flow-path option for the fluid. The multiple envelope moves to illustrate the change of flow paths when the valving element inside the component is shifted to its finite positions.
3. Multiple-envelope valves capable of infinite positioning between certain limits are symbolized as above with the addition of horizontal bars drawn parallel to the envelope. The horizontal bars are clues to the infinite positioning function of the valve represented.

Single and multiple *envelopes*.

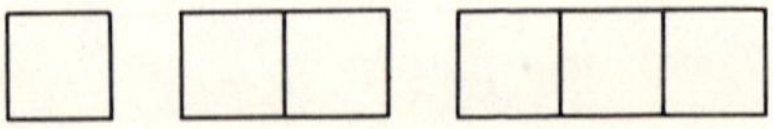

Envelopes with *ports*.

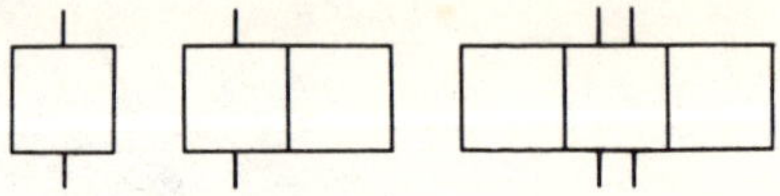

Ports *blocked internally* (symbol systems No. 1 and No. 2).

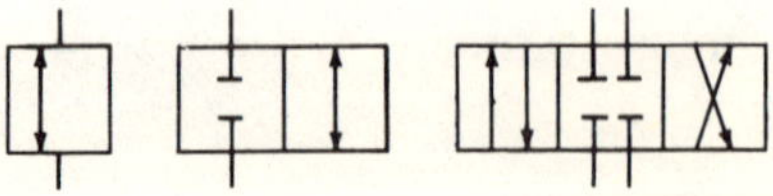

Flow paths *internally open* (symbol systems No. 1 and No. 2).

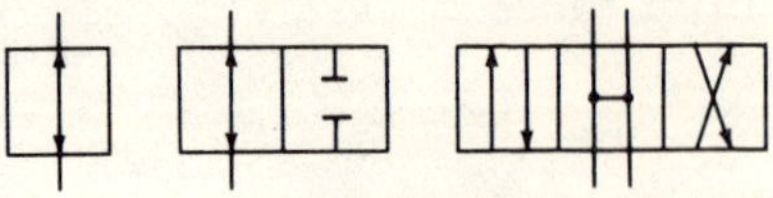

Flow paths *internally open* (symbol system No. 3).

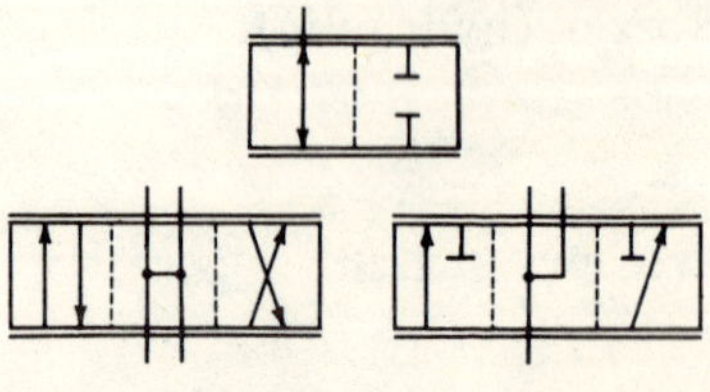

Two-Way Valves (Two-Ported)

Manual shutoff *on-off* valve in *off* position (upper) and in *on* position (lower).

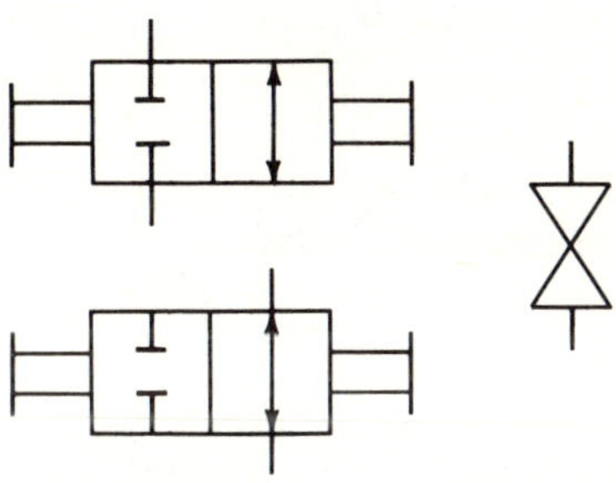

Check (composite symbol) in which flow to the right-hand side is blocked, but flow to the left-hand side is permitted.

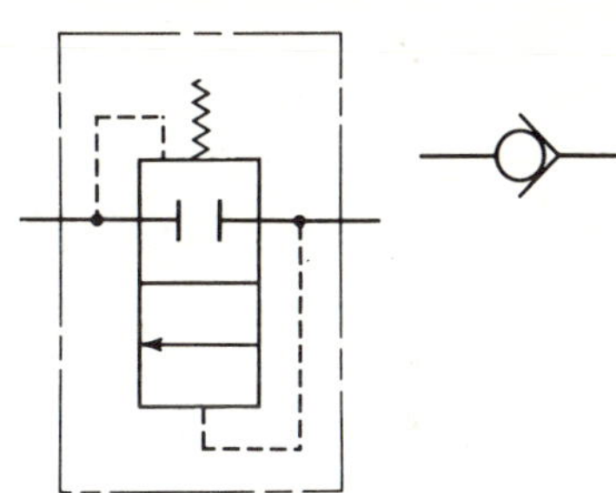

Check, *pilot-operated to open.*

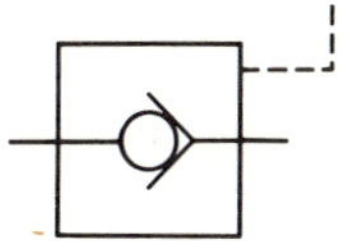

Check, pilot-operated *to close.*

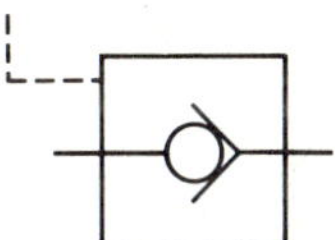

Normally closed (left) and *normally open* (right) *two-position* two-way valves.

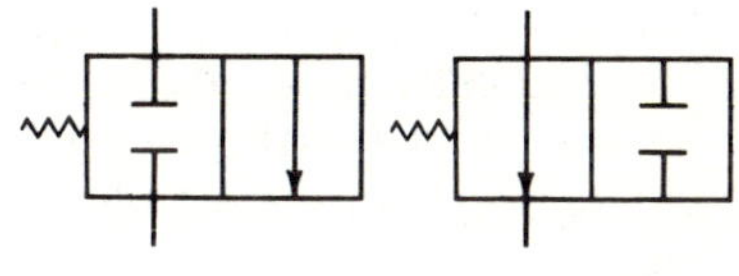

Infinite position *normally closed* (left) and *normally open* (right) two-way valves.

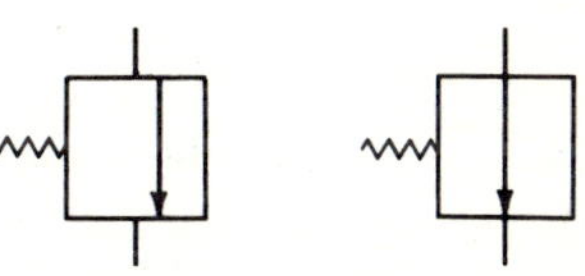

Three-Way Valves

Normally open two-position three-way valve.

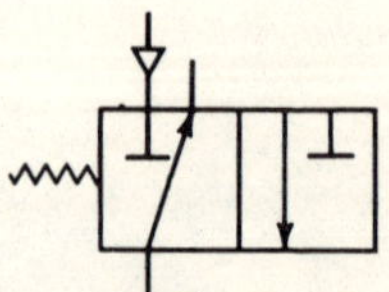

Normally closed two-position three-way valve.

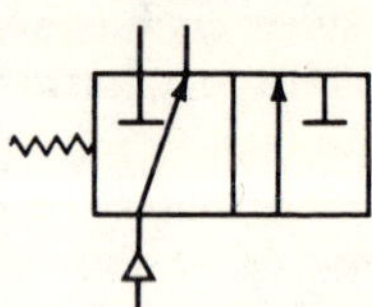

Distributor valve. Pressure is distributed first to one port and then to the other port.

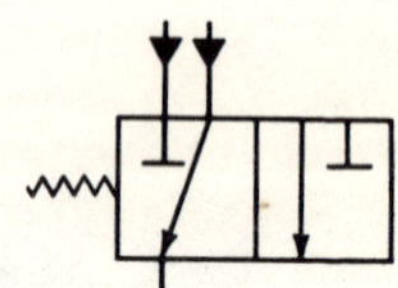

Two-pressure two-position three-way valve.

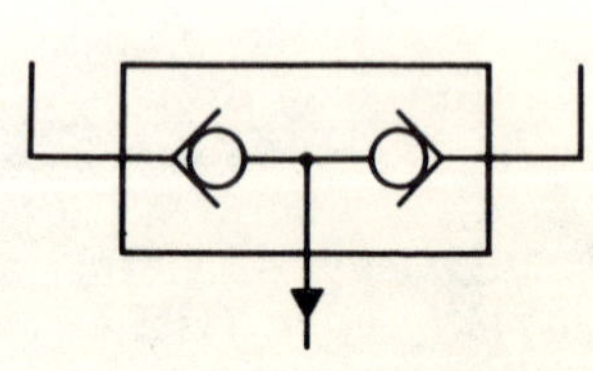

Double check valve *without cross bleed* (one-way flow). Valves with two poppets usually do not permit pressure to momentarily cross bleed to return during transition. Valves with one poppet may permit cross bleed.

Double check valve *with cross bleed* (reverse flow permitted).

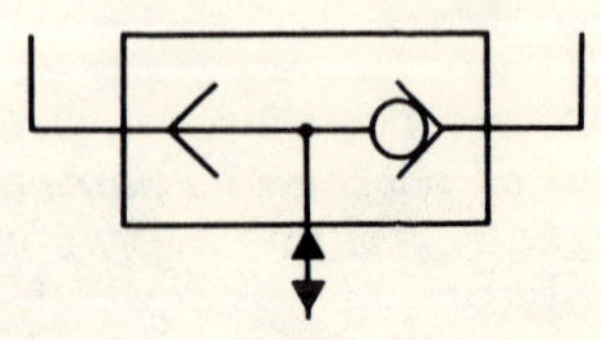

Four-Way Valves

Snap action with transition two-position four-way valve. As valve element shifts positions, it passes through an intermediate position. If this "in-transit" condition is essential to circuit function, it can be shown in the center position, enclosed by broken lines.

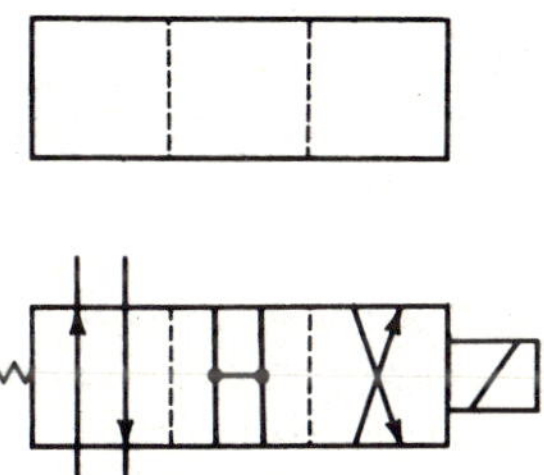

Typical flow paths for *center position* of three-position four-way valves.

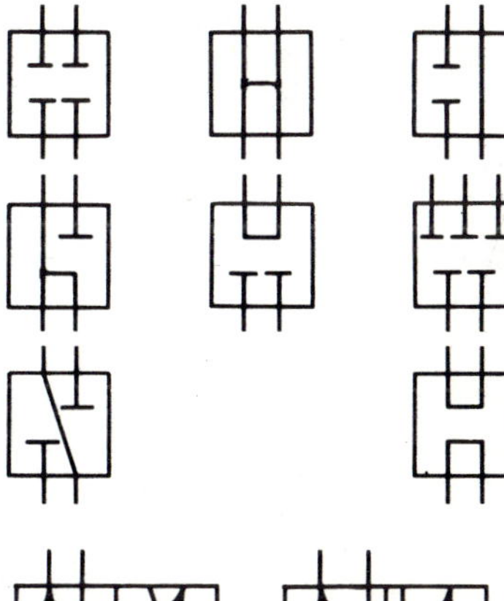

Two-position four-way valve in *normal* position.

Two-position four-way valve in *actuated* position.

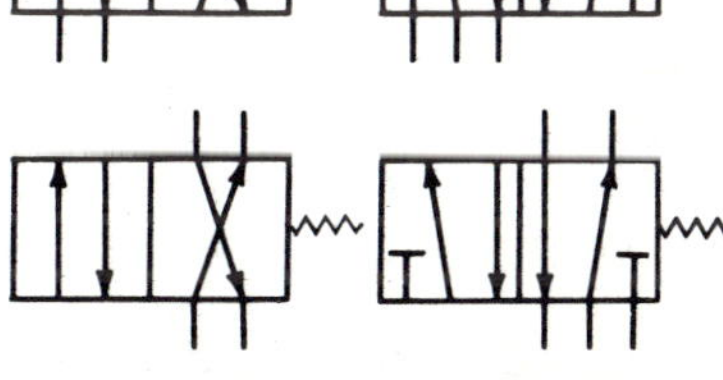

Three-position four-way valve in *neutral* position.

Three-position four-way valve in *actuated left-hand* position.

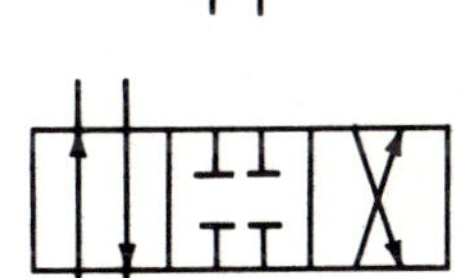

Three-position four-way valve in *actuated right-hand* position.

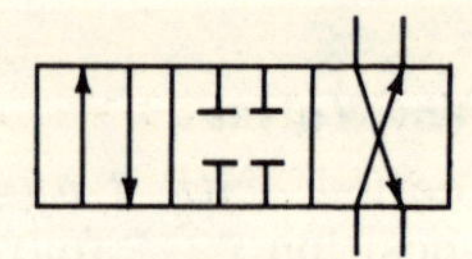

Infinite Positioning (Between Open and Closed)

Normally closed valve (left) and *normally open* valve (right).

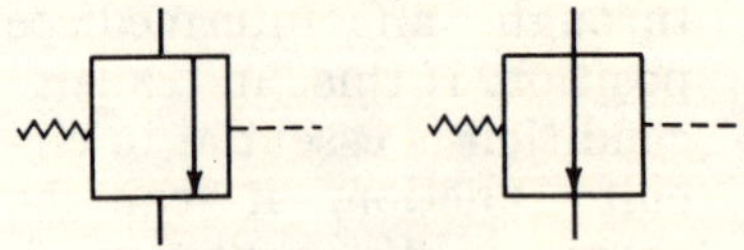

Pressure Control Valves

Pressure relief valve (simplified symbol) in *normal* position (left) and in *actuated* (relieving) position (right).

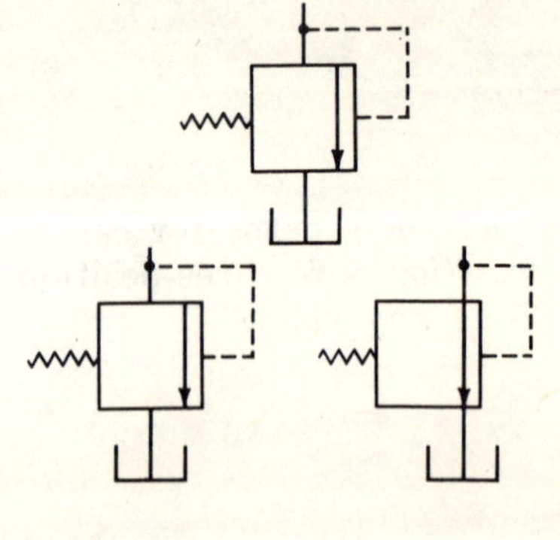

Sequence valve.

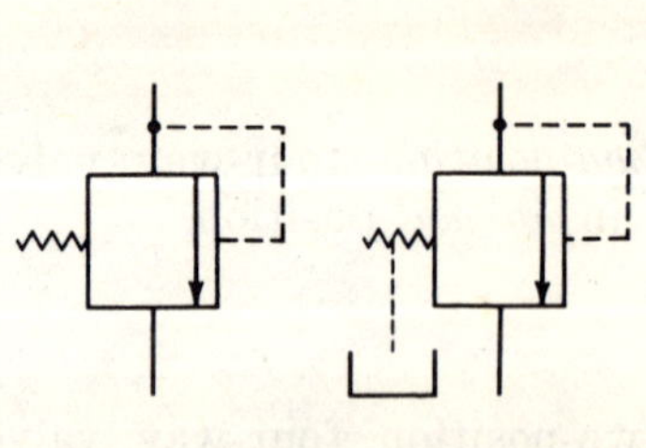

Pressure-reducing valve.

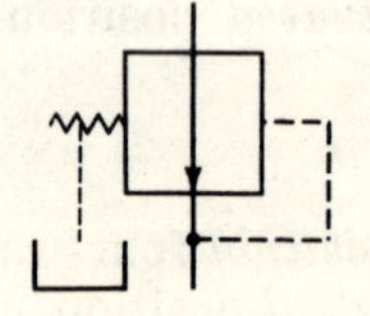

Pressure-reducing and relieving valve.

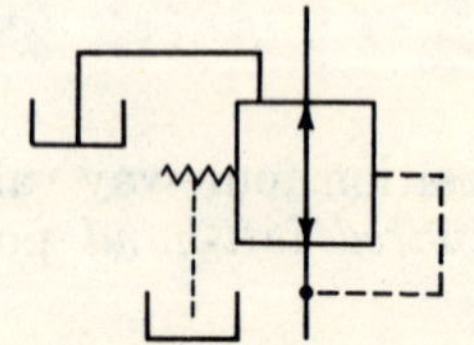

Infinite Positioning Valves

Three-way valves.

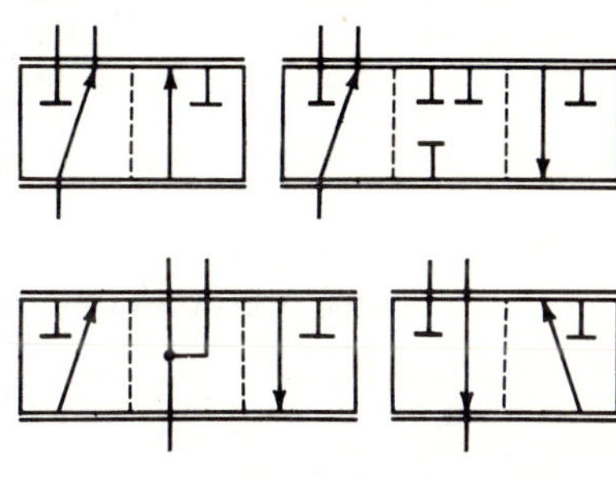

Four-way valves.

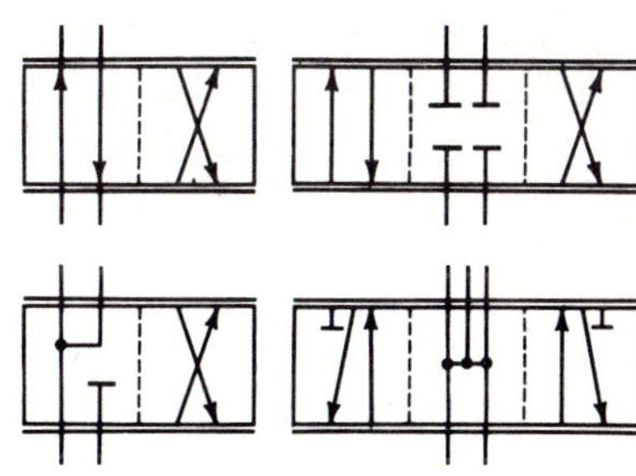

Flow Control Valves

Adjustable *with bypass* flow control. Flow is controlled in the right-hand direction. Flow in the left-hand direction bypasses the control.

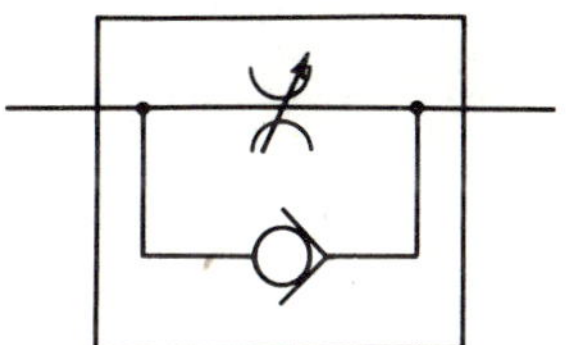

Noncompensated (flow control in each direction) *adjustable* flow control valve.

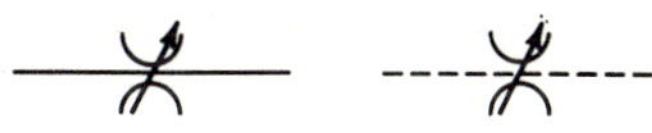

Adjustable and *pressure-compensated with bypass* flow control.

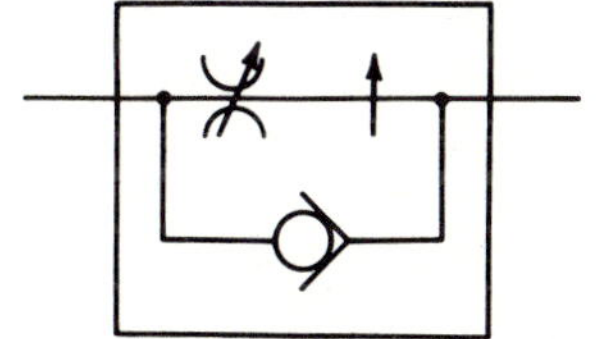

Temperature- and pressure-compensated adjustable flow control.

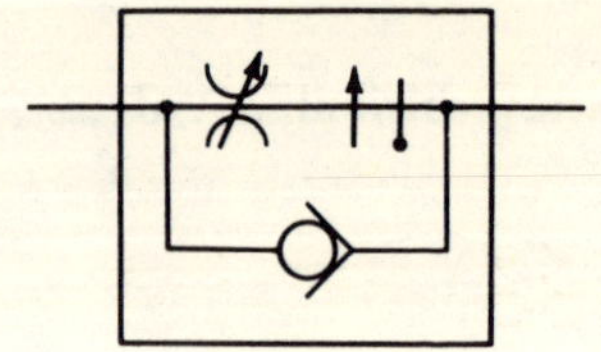

REPRESENTATIVE COMPOSITE SYMBOLS

Component enclosure. The enclosure may surround a complete symbol or a group of symbols to represent an assembly, and it is used to convey more information about component connections and functions. The enclosure indicates the extremity of the component or assembly. External ports are assumed to be on the enclosure line and indicate connections to the component. The flow lines cross the enclosure line without loops or dots.

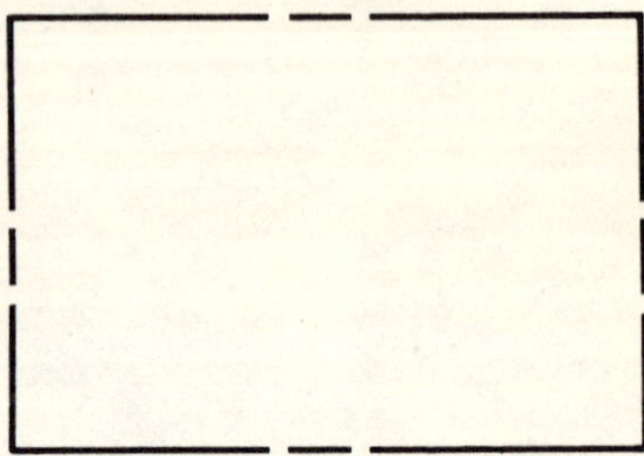

One inlet and two outlets fixed-displacement *double pumps.*

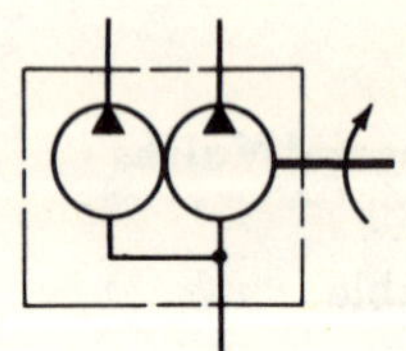

Pump with integral *variable-flow rate control* with overload relief.

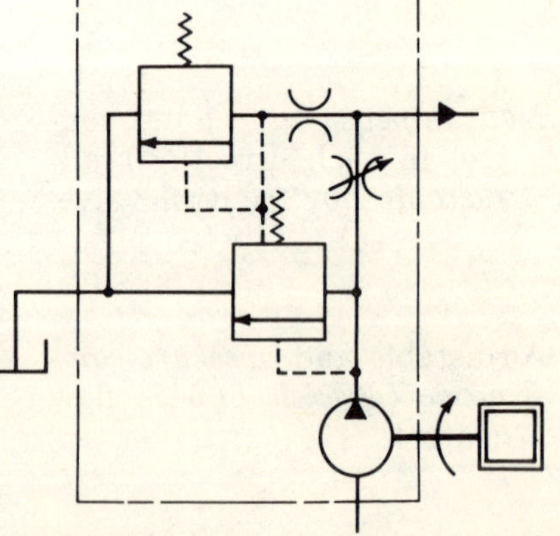

Double pumps with integral *check unloading* and two outlets.

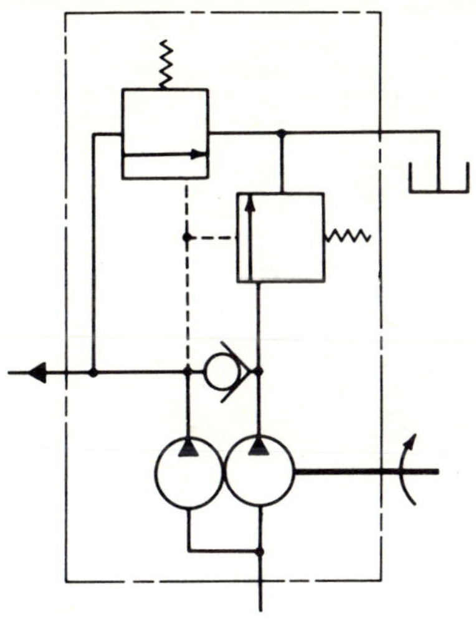

Variable-displacement pump with integral *replenishing* pump and control valves.

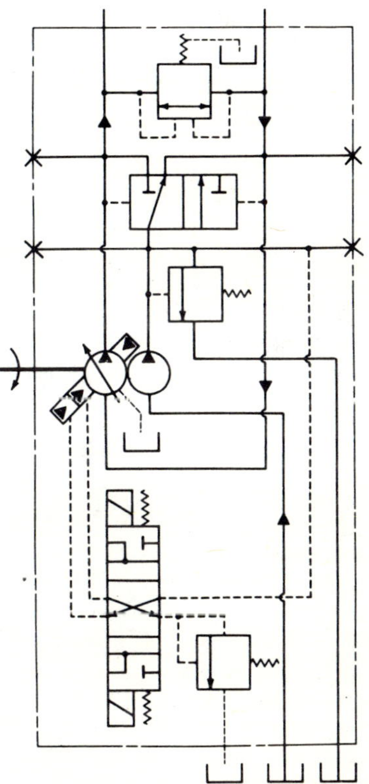

Balanced-type *relief* valve.

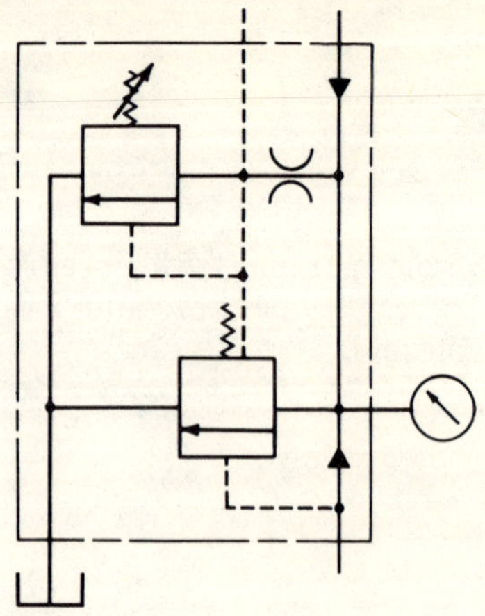

Remote-operated *sequence* valve with integral check.

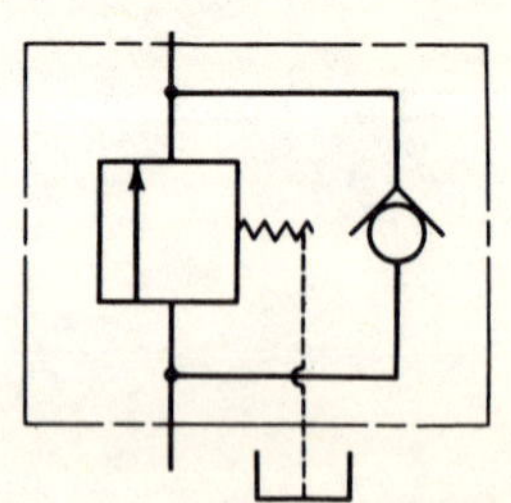

Remote and direct-operated *sequence* valve with differential areas and integral check.

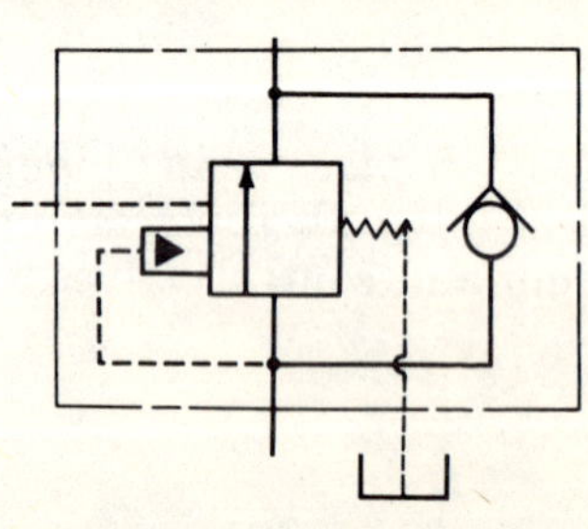

Pressure-reducing valve with integral check.

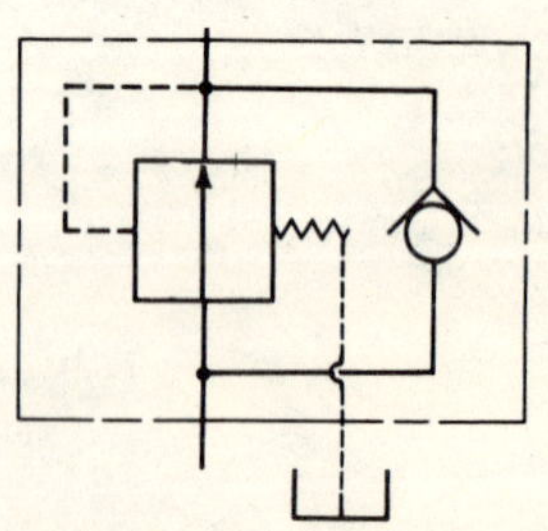

Multiple three-position manual *directional control valve* with integral check and relief valves.

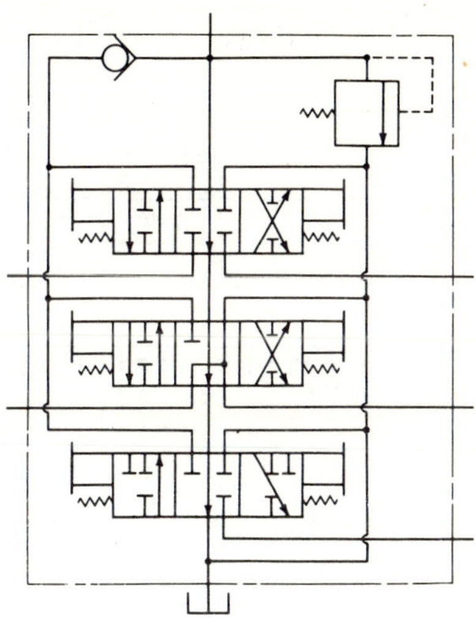

Differential *pilot-opened* check valve.

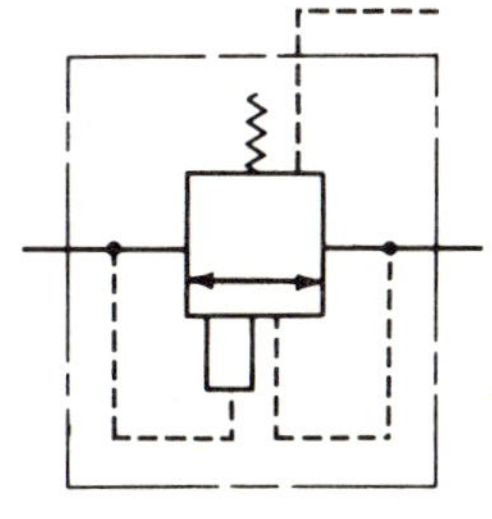

Differential *pilot-opened and pilot-closed* check valve.

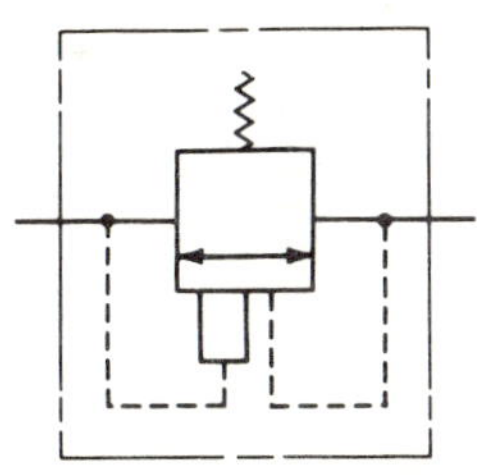

Variable *pressure-compensated* flow control and overload relief valve.

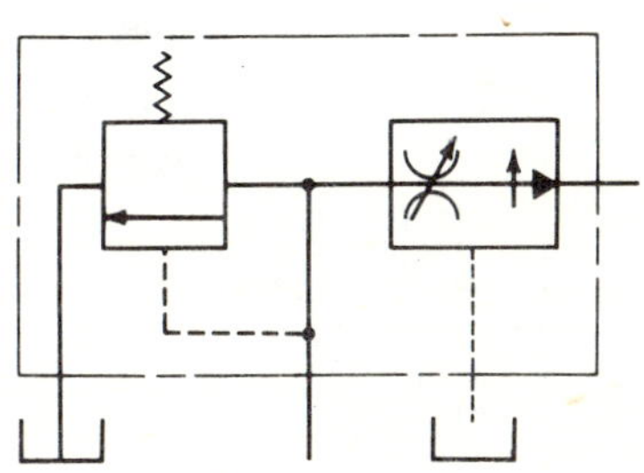

Five-position *cycle control panel.*

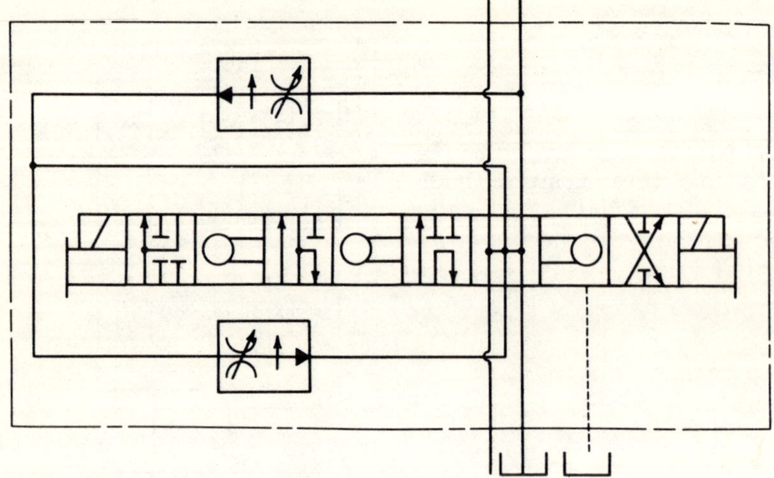

Panel-mounted separate units furnished as a *package* (relief, two four-way, two check, and flow rate valves.)

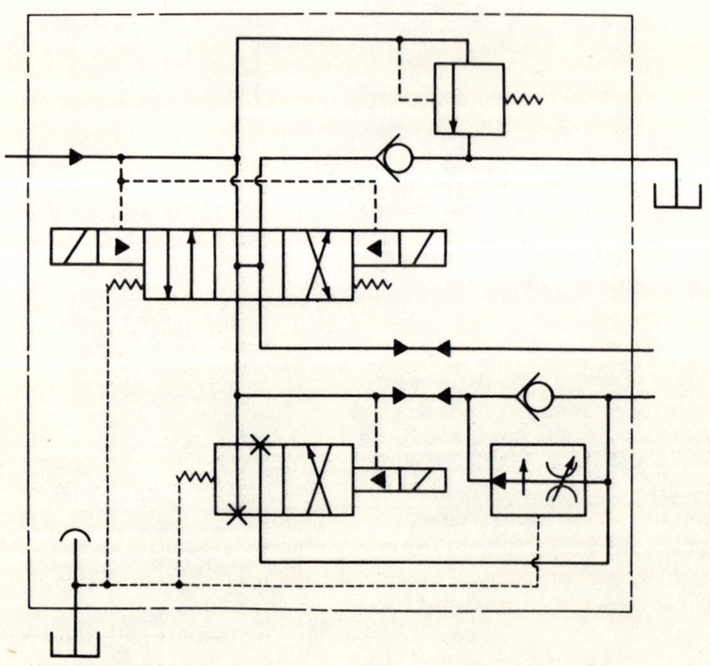

Variable-displacement *pump motor* with manual, electric, pilot, and servo control.

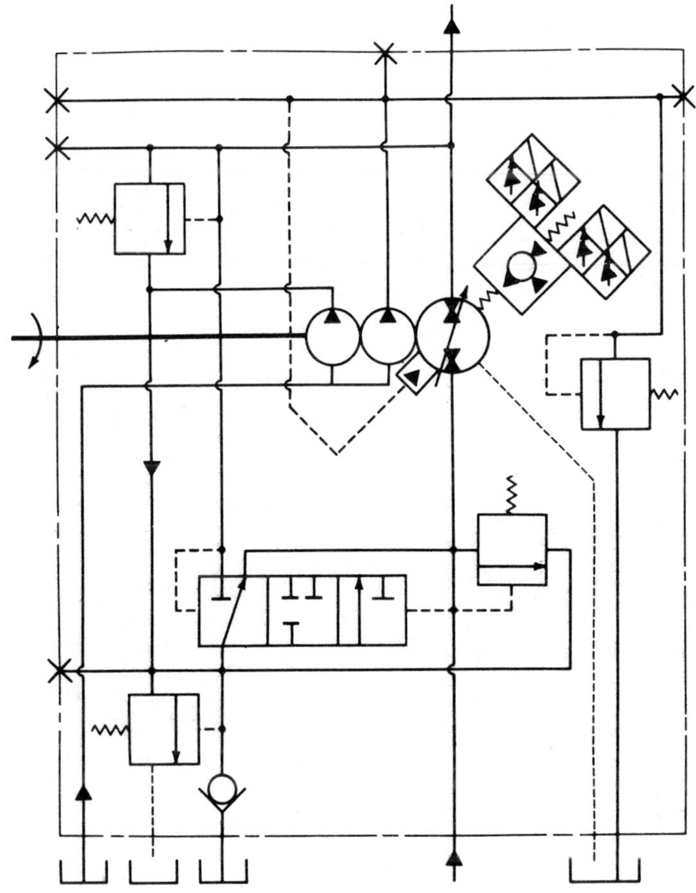

COLOR OR PATTERN CODE

The interconnecting lines in cutaway diagrams are sometimes colored or patterned to indicate pressure, flow, special functions, and different fluids during various phases of operation. The code for each condition used should be identified and illustrated by a note, preferably located in the lower left-hand corner of the sheet. Only those lines which perform active functions for the phase shown should be coded. If a condition fits more than one function, the code which provides the greatest emphasis and clarity should be selected. When applicable, the following color and pattern code should be used.

FUNCTION	COLOR	PATTERN
INTENSIFIED PRESSURE	VIOLET	
SUPPLY PRESSURE	RED	
CHARGING PRESSURE REDUCED PRESSURE PILOT PRESSURE	ORANGE	
METERED FLOW	YELLOW	
EXHAUST	BLUE	
INTAKE DRAIN	GREEN	
INACTIVE	BLANK	

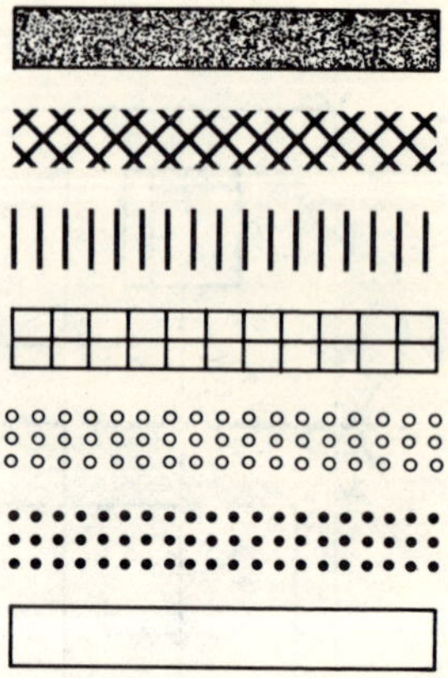

CHAPTER 2

Power Unit Basic Circuits

Work is accomplished in a hydraulic system by the use of a liquid as a carrier of energy from its generating source to its point, or points, of application. The properties of a good hydraulic fluid that make it an ideal carrying agent are:

1. Comparative incompressibility
2. Adaptability to an infinite variety of volumetric configurations
3. Capacity for a wide range of movement of flow rates
4. Capacity for energy in the visable form of pressure

The two parameters that determine the work capacity of any hydraulic system are flow (gpm) and pressure capability (psi). The flow generator, or hydraulic pump, is a volumetric displacement component that is designed to generate a rate of liquid flow against a resistance. The measure of the flow rate is usually read in terms of gallons of liquid displaced per minute, or gpm. The magnitude of the flow resistance is usually read in terms of force/area, or pounds per square inch (psi).

The hydraulic pump with its supplementary accessories comprises the energy-generating unit of a hydraulic system. This combination of pump and supplementary accessories is commonly labeled as the "hydraulic power unit."

The number of components contained in a hydraulic power unit may vary, depending upon the complexity of the circuit being supplied and the demands being made upon it. Such a power unit may include some or all of the following:

1. A volumetric displacement unit (pump)
2. A reservoir for liquid storage
3. A motive force for the pump
4. A device for limiting the maximum pressure resistance
5. A unit for capturing liquid contaminants
6. A unit for reading pressure resistance
7. A unit for heating the liquid
8. A unit for removing heat from the liquid
9. Interconnecting plumbing

In addition to these items, functional control components of the working circuit may be mounted directly on the power-unit package. When directional control valves are mounted in this manner, they are considered to be a part of the hydraulic power-unit package.

The pump may be the type that generates a given, or fixed, rate of flow, or it may have the capability of variable volumetric displacement. The form the pump takes should be indicated in the circuit by employing the appropriate symbol in the circuit diagram.

The reservoir may be required to serve many purposes and should be designed accordingly. It may not only provide for fluid storage, but also effect some degree of heat dissipation and contaminant control. It usually serves, also, as a platform for mounting auxiliary components. For the purposes of diagramming the circuit, however, the proper practice is to use the simple reservoir symbol, shown in Chapter 1. The number of auxiliary components contained in the total hydraulic power-unit package is indicated by including all such component symbols within a package envelope enclosure.

Fig. 2-1 depicts a simple hydraulic power-unit package employing reservoir R, suction filter B, fixed displacement pump P, drive motor M, relief valve A, pressure gage G and shutoff valve V, each shown in its proper relationship to the others. These positions are functional positions only and are not intended to depict any dimensional information or actual physical locations. It should be noted that two reservoir symbols are depicted in this circuit. This duplicity does not signify more than one physical reservoir, but is merely a convenient method of simplification for the designer. Whenever the circuit must provide multiple connections to return fluid to the reservoir, the use of additional reservoir symbols eliminates the necessity of drawing long lines to indicate such returns. The envelope en-

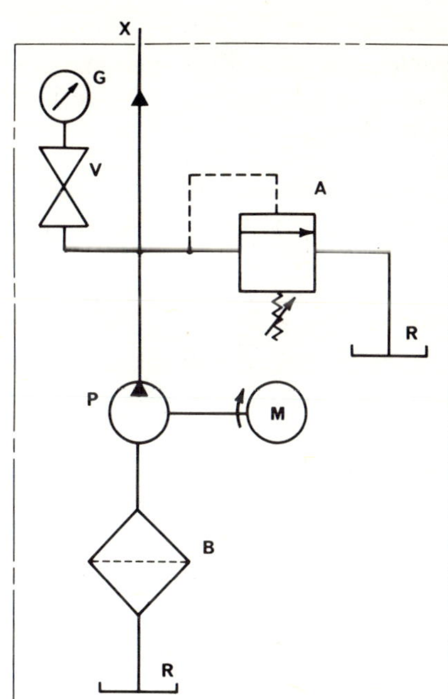

Fig. 2-1. Simple hydraulic power-unit package.

closure symbol includes all of the components employed in the package, with line X indicating the pressure-supply connection external to the package.

The electric motor drive, depicted in Fig. 2-1, is most commonly employed in industrial applications. Other drive units that may be used are air motors, internal combustion engines (sometimes with gear reducers), and power take-offs (frequently found in mobile applications). The drive method used should be incorporated into the circuit diagram by use of the appropriate symbol.

FILTERING

Fig. 2-1 depicts the use of a filter unit in the supply, or "suction," line to the pump. When used in this location, the unit may take either of two forms. The simplest form is a nonencapsulated screen that is submerged in the reservoir well below the minimum fluid level. Such a unit is usually labeled "suction strainer" and requires entering the fluid for servicing or replacement. The alternate form is designed for external mounting to the reservoir for greater convenience in servicing. This latter form provides an encapsulating enclosure for the filter element, with plumbing connections provided in the body of the enclosure. The filter element can usually be removed without breaking any conduit connections. The symbol employed in the circuit design is the same for either form, except that an enclosure box may be used to indicate the encapsulated form. The degree of filtration may be indicated on the circuit design, particularly if this factor is critical to the proper functioning of the circuit. A selection of screen sizes is available for suction filters, but filtration below about 70 microns is usually not recommended. If in doubt about a particular choice, pump manufacturers' specifications should be consulted.

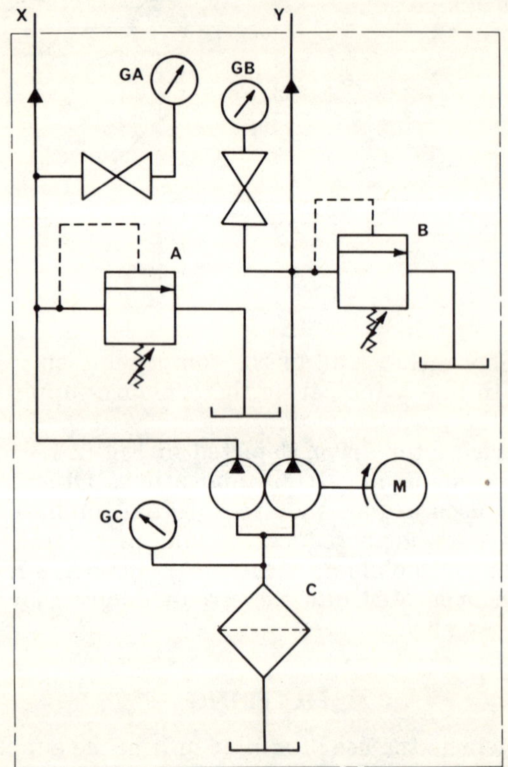

Fig. 2-2. Package using double pump with single filter.

Fig. 2-2 depicts a hydraulic power-unit package incorporating a double pump with a single suction filter common to both pump elements. The filter must be of adequate capacity to handle the combined flow of the two elements. Although the suction lines pass through the common filter, the pressure outputs of the two elements are independent of each other. The two separate systems may be of unequal volume (gpm), and the maximum pressure values may be different. The maximum pressure value of the system on the A side of the compound circuit is established by relief valve A, with gage GA sensing its pressure resistance. The maximum pressure value of the B side element is established by relief valve B, with gage GB sensing its pressure resistance. Gage GC is a vacuum-sensing unit used as an indication of the flow resistance through filter C. As the filter element becomes loaded with contaminants, the resulting increase in pressure drop will appear as an increase in vacuum

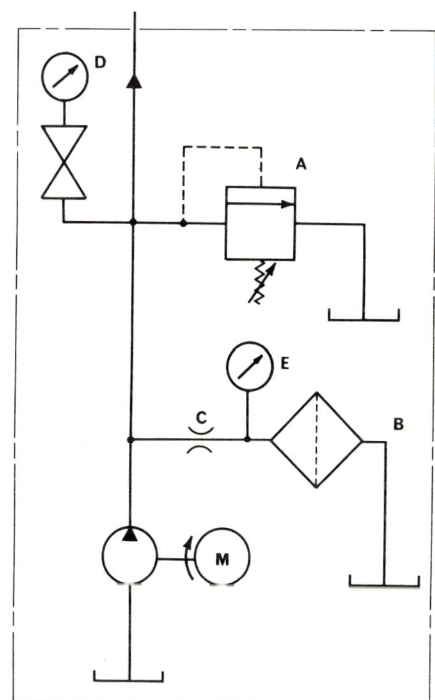

Fig. 2-3. Bypass filtering.

indication on the gage. While many encapsulated-type suction filters incorporate an indicating device for this purpose, a simple vacuum gage, used as indicated, will perform the same function, even when a submerged-type suction strainer is employed.

The circuit designer has a variety of filtering options to choose from. One such option is the bypass or partial filtration technique depicted in Fig. 2-3. Only a portion of the pump displacement passes through filter B, but it is reasonable to assume that all of the fluid in the system will eventually pass through the B filter element. The portion of the pump displacement that is permitted to pass through the filter is determined by orifice C. Since the volume passing through the filter is not available to perform work elsewhere in the system, this volumetric loss should be taken into account in sizing the pump. Although this volumetric loss appears to be inefficient, such an arrangement permits much finer filtration than can be accomplished with the suction-type filter.

The filter employed in this diagram should never be subjected to any appreciable pressure due to the flow-limiting function of orifice C. Gage E is a pressure-sensing unit that will indicate increasing back pressure as the filter element becomes loaded with contaminants. If maintenance has been inadequate and the filter element is neglected, excessive loss of flow capacity could result in maximum system pressure values within the filter case. For this reason, a filter with an adequate pressure rating should be employed. If excessive contaminants are permitted to enter the reservoir from other sources, a suction filter should also be employed to provide some degree of protection for the pump.

The bypass filter employed in Fig. 2-3 may be of the low-pressure type if it is protected by a pressure-reducing valve as shown in Fig. 2-4. The volume passing through filter C is determined by orifice D, and the maximum pressure that can be created within the C filter case is determined by the setting of pressure-reducing valve A. It should be noted that the drain connection from pressure-reducing valve A terminates in the reservoir *above* fluid level.

Fig. 2-5 offers an alternate method of partial filtration. Variable orifice B limits the volume of fluid that can return directly to the reservoir from the tank line of directional control valve A. Excessive return flow is diverted through filter C. No pump volume is lost to the working portion of the circuit. However, any increase in flow resistance through the filter element will result in a loss of that pressure to perform work. The presence of directional control valve A within the enclosure symbol indicates that the control valve is attached to, and is a part of, the power-unit package.

Any hydraulic system that requires extremely fine metering cannot tolerate the pressure of contaminating particles. Such particles will cause a "silting" action across small orifices, as these orifices begin to act as filters. This problem becomes particularly acute when servo valves are employed. Some servo-

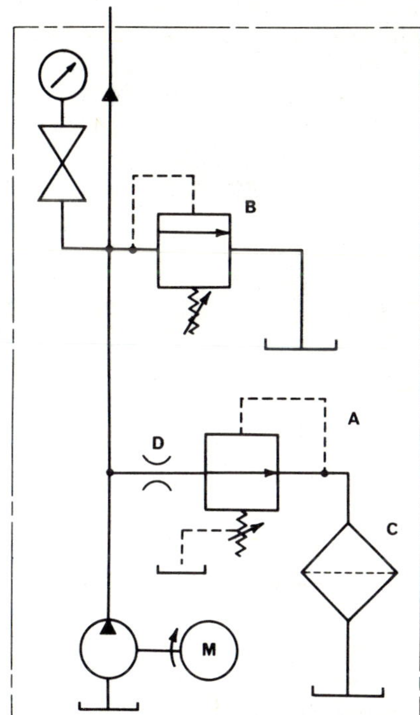

Fig. 2-4. Low-pressure bypass filtering.

valve specifications call for 10-, 5-, or even 3-micron filtration to ensure proper functioning. Filter units to provide such fine particle removal are readily available as pressure-line filters.

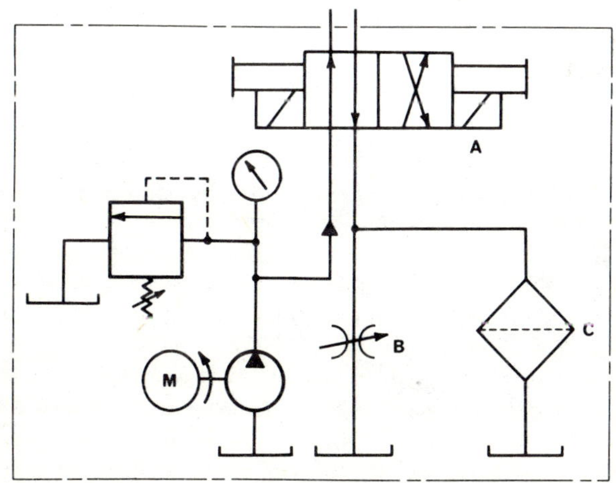

Fig. 2-5. Alternate method of partial filtering.

It is usually wise and economical to preceed such filters with coarser mesh units for capturing the larger contaminating particles. This practice will extend the life of the low-micron elements and reduce the frequency of having to service the filter system. Fig. 2-6 depicts a dual-filter circuit that matches this requirement.

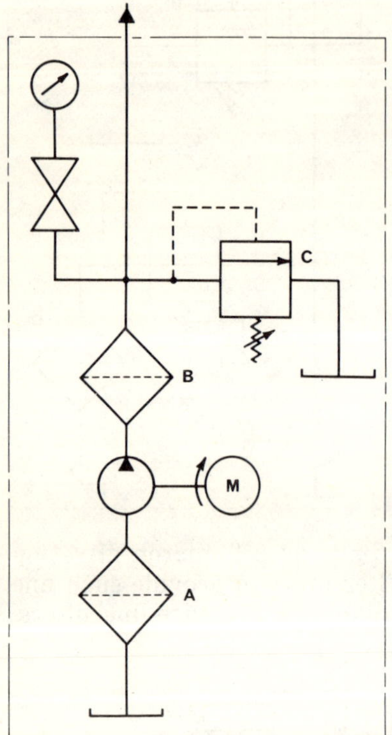

Fig. 2-6. Dual filtering circuit.

Filter A could be a relatively inexpensive suction filter of 200-mesh screen that will capture most particulate materials of 40-micron size or larger. It also affords some protection to the pump. Filter B is a low-micron, pressure-type filter with a low-micron filtering element. Once the fluid in such a system has been "cleaned," it should remain clean as long as the system is properly protected against entry of outside contaminants.

Another filtering option available to the circuit designer is return-line filtration (Fig. 2-7). Check valve A has a cracking pressure of 65 psi. Filter D should be of sufficient capacity to handle the full return line flow from valve E. If oversize piston rods are present in the operating cylinders, the increased ex-

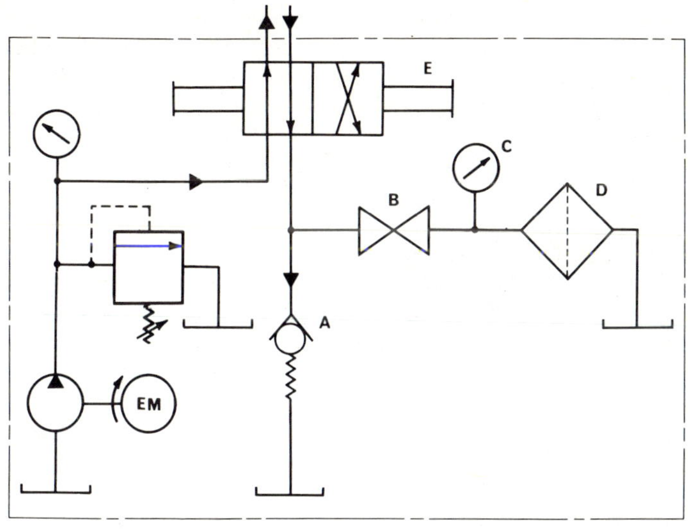

Fig. 2-7. Return-line filtering.

haust flow rate from the cap-end cylinder chambers should be taken into account. Shutoff valve B can be closed for removal and maintenance of the filter element while the power unit is in operation. Gage C is a pressure-sensing unit that will indicate the condition of the flow capacity of the filter element. When the element becomes loaded with sufficient contaminant particles to create 65 psi back pressure (flow resistance), the gage will so indicate. As this pressure is reached, check valve A opens, permitting flow directly to reservoir.

The potential pressure buildup, cited in reference to Fig. 2-3, can be avoided by the use of an encapsulated filter with a built-

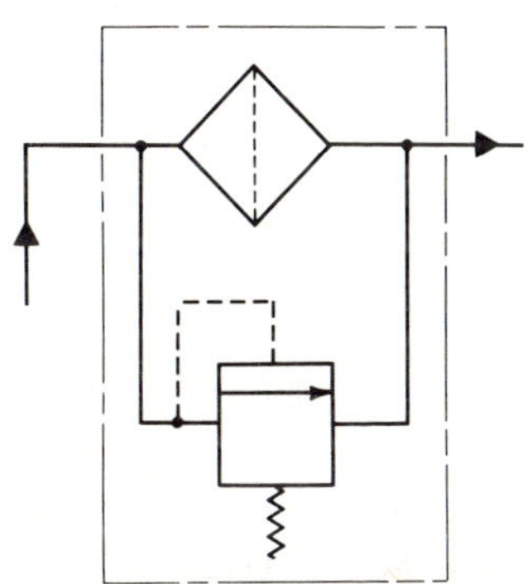

Fig. 2-8. Filter with built-in relief valve.

in relief valve. Such a filter is depicted by the symbol shown in Fig. 2-8. If the filter element becomes sufficiently loaded with contaminants to generate excessive flow resistance, the built-in relief valve opens to prevent overpressuring the case. When used as a suction line filter, the opening of the relief valve prevents starvation of the pump.

TEMPERATURE CONTROL

The greatest efficiency can be maintained in a hydraulic system if the fluid temperature is kept within a range of 100 to 140 °F. If the fluid temperature is too low, the viscosity increases with a resulting increase of flow resistance. Input hp is thereby wasted in merely overcoming this flow resistance. If the system is to be subjected to low ambient temperatures, emersion-type heaters should be installed in the fluid reservoir to bring the fluid up to an efficient operating temperature. To avoid overheating, such heaters should be equipped with a thermostatic control. If the system is required to stand idle while exposed to weather or other chilling atmospheres, a timed control of the heater would be in order. Such an arrangement

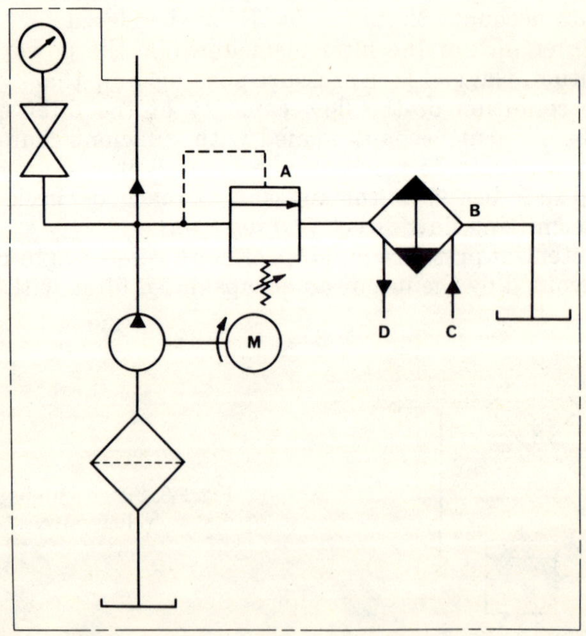

Fig 2-9. Hydraulic power-unit package with heat exchanger.

would provide preheating of the fluid automatically, in advance of the beginning of a work shift.

The most frequently encountered problem of temperature is the buildup of heat within industrial hydraulic systems. Excessive heat not only causes hydraulic fluids to deteriorate, but it also changes the characteristics of the fluid and the performance of the components within the system. Uncontrolled temperatures within a hydraulic system can either destroy or render ineffective many of the standard materials that are

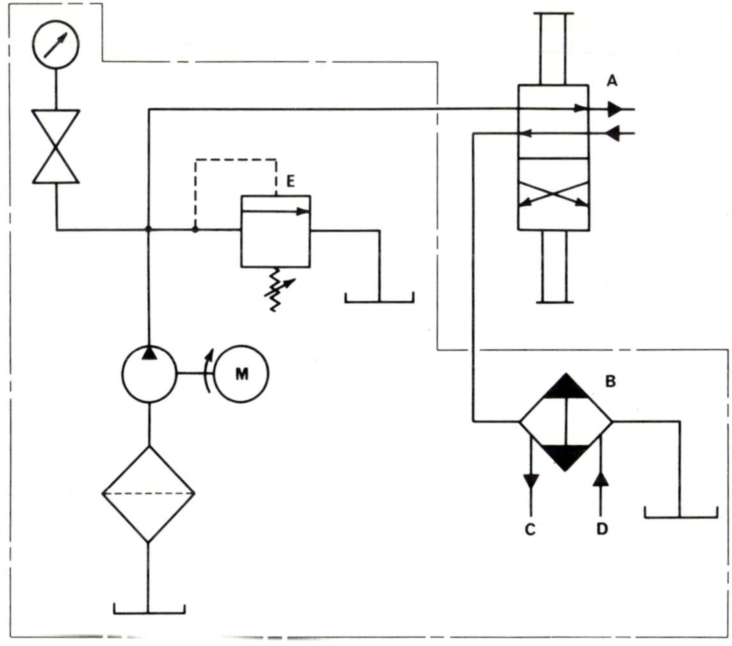

Fig. 2-10. Heat exchanger installed in tank line.

used in system components. Clearances of moving or mating parts made of dissimilar materials may be altered due to different coefficients of expansion. These changes may create excessive clearance flows or increased friction resistance resulting in thc creation of additional heat.

One of the properties or characteristics of a hydraulic oil that helps to determine its performance within a hydraulic system is its viscosity, or resistance to flow. The viscosity of an oil changes as its temperature changes. Since the inherent resistance to the flow of a fluid determines its rate of flow though a given orifice, any unwarranted temperature change can result in varying rates of movement in the hydraulic components being operated. Thus, a system or hydraulically operated machine that is adjusted for a given speed at the beginning of a work day may alter its performance rate to such an extent that numerous adjustments are required during an operating period.

Standard seal materials that are subjected to temperatures in excess of their intended operating range may "cure out," or harden, rendering them ineffective or useless. Hydraulic com-

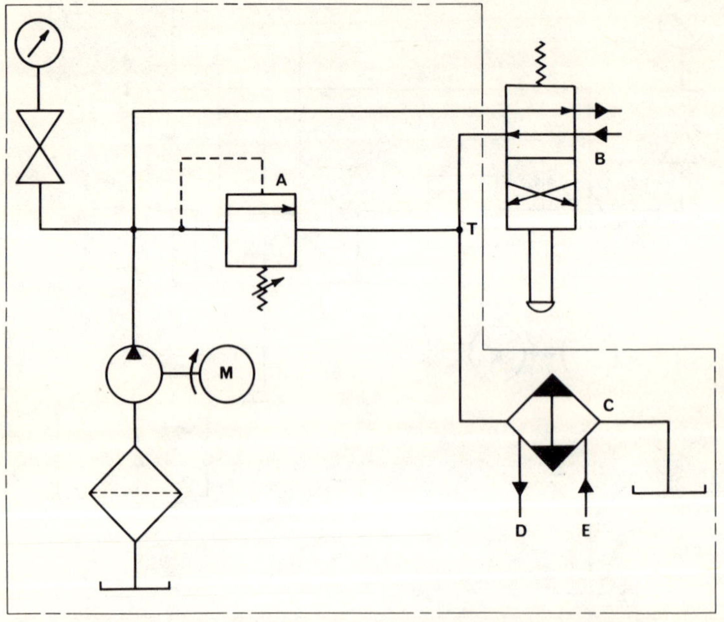

Fig. 2-11. Heat exchanger cools fluid from relief valve and tank.

ponents can be equipped with seals that are made from compounds capable of withstanding temperatures in excess of 500 °F. These seals, however, are much more expensive than the standard seals, and they are only a partial answer to a problem that may be handled more easily by controlling the temperature of the fluid.

Most of the heat that is encountered in a hydraulic system originates at the pumping unit. Heat is induced into the hydraulic fluid as a direct result of the work that is performed on the fluid in moving it against a pressure resistance. The amount of heat generated will show a positive relationship to the amount of input horsepower that is wasted. The tendency of a fluid to hold its heat results in a pyramiding of heat accumulation to a point where it becomes a problem within the hydraulic system.

The most common remedy for excessive temperatures in a hydraulic system is the heat exchanger. The two main types of heat exchanger used in industry are the air-to-oil type using air as the heat transfer agent, and the water-to-oil type, which transfers the heat to water.

Fig. 2-9 depicts a hydraulic power-unit package that incorporates a heat exchanger using water as the cooling medium.

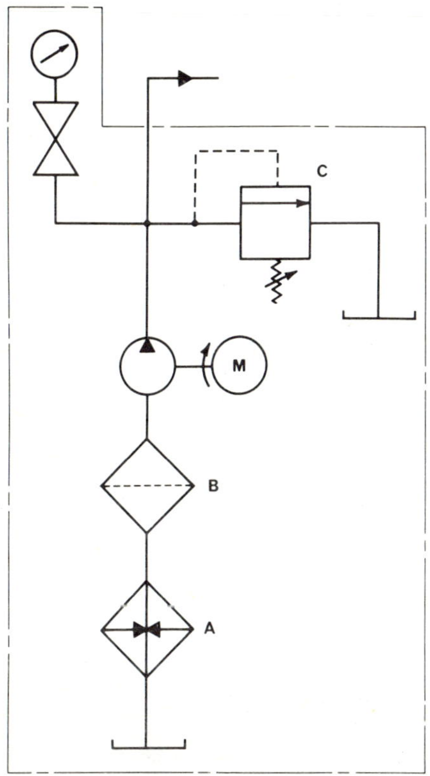

Fig. 2-12. Low-pressure heater installed in suction line.

Cooling unit B is installed in the tank line from relief valve A. If the operating circuit permits fluid to spill over the relief valve for an appreciable percentage of the total operating time, adequate heat removal should be obtained.

For those circuits where very little relief-valve spillage is encountered, Fig. 2-10 will offer a superior cooling effect. Maximum exposure of the fluid to the cooling medium, under varying operating conditions, will be obtained by the use of the circuit depicted in Fig. 2-11. Here, both the fluid from the relief valve, as well as the fluid from the working circuit, are joined at junction T and channeled through heat exchanger C. Wherever possible, all returning oil, except for that returned in the drain lines, should be channeled through the heat-exchanger unit. Most heat-exchanger units are designed for the low pressures that are encountered in the return lines, but the high-

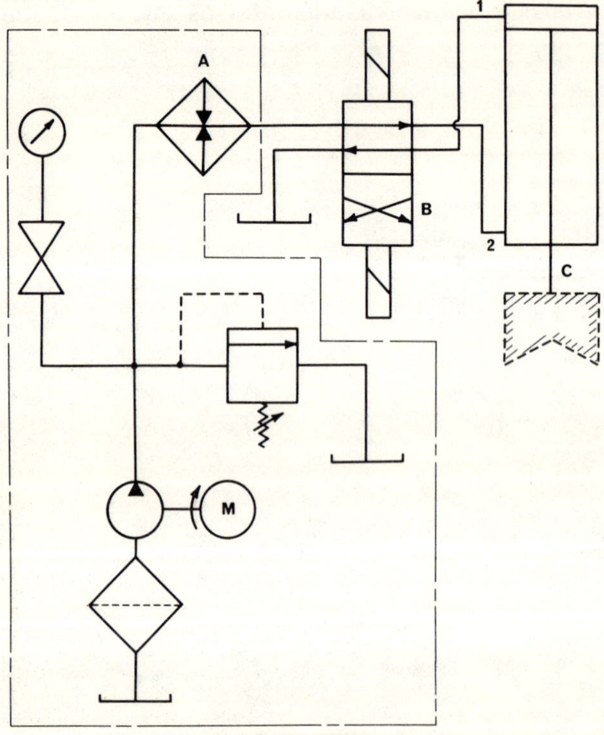

Fig. 2-13. Heater installed in pressure line.

pressure units are available, at much greater cost. Occasionally, a condition may exist where it is imperative to install the cooling unit in the pressure line, but these conditions are rather uncommon. Other methods of combating heat will be discussed in Chapter 8.

Figs. 2-12 and 2-13 depict the use of heat-inducing units for raising the temperature of the operating fluid. Heater A in Fig. 2-12 is a low-pressure unit installed in the reservoir or suction line and pressure matching is not particularly critical. Heater

A, in 2-13, is shown in the pressure line and must be carefully selected to match the maximum pressure rating of the system due to the obvious safety considerations involved with high-pressure fluid at a high temperature.

CHAPTER 3

Press Circuits

The functional requirements of a hydraulic press circuit may span a wide range of possible specifications. The examples illustrated herein are not exhaustive, but are intended to suggest a variety of control combinations that may be employed as shown, or in combinations. Although these circuits are labeled as press circuits, the control techniques used are basic and can readily be applied to other work requirements. The final design of any circuit should be predicated on the performance desired by the designer.

The desired performance of a press circuit may usually be achieved in a variety of ways and by selecting from a number of component combinations. The wise designer will keep the circuit as simple as is practicable, while attempting to anticipate all potential functional problems the work requirement might suggest.

MANUAL OPERATION

The simplest press circuit provides for a straightforward close and open movement of the press ram. Fig. 3-1 depicts such an action employing two-position, spring-offset, manually operated, three way, directional control V controlling a single-acting, weight-return, up-acting press ram. The force of the press ram is a product of fluid pressure multiplied by the piston area of the cylinder. The speed of closure is determined by the volumetric displacement rate of the pump. The speed of opening is determined by the combined weight of the ram, platen, and work fixture in relation to the return flow path of the fluid to reservoir. The maximum pressure (psi) is established by the setting of relief valve R. Functionally, the circuit provides two operational conditions: (1) platen retracted position as shown, and (2) platen raised or exerting force upward when control valve V is held manually actuated.

Fig. 3-2 illustrates a variation of Fig. 3-1 by utilizing a double-acting cylinder and four-way, two-position, manually operated, spring-offset control valve V. The functional characteristics are the same, except this press is a down-acting unit that utilizes hydraulic pressure to return the piston and attached platen. The hydraulic power supply is required to hold the piston and platen in the up position.

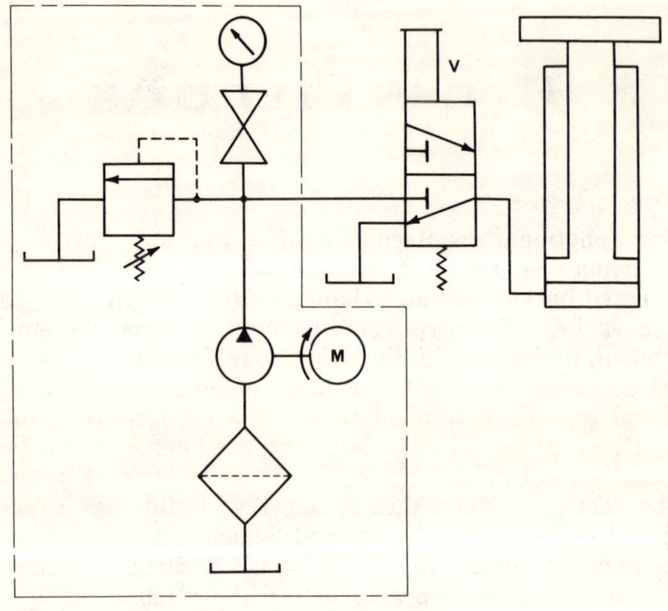

Fig. 3-1. Simple manually operated press circuit.

INCHING

An additional operating characteristic often required is the capability of inching the press platen or ram and achieving hydraulic stability in one or more intermediate positions. This characteristic is readily achieved in Fig. 3-3 by substituting three-position, closed-center, four-way directional control valve V in place of the two-position valve employed in Fig. 3-2. This

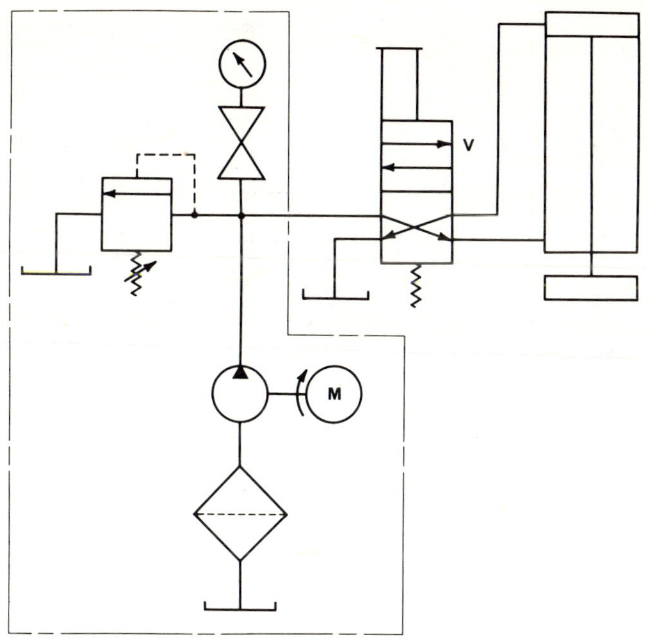

Fig. 3-2. Press circuit requiring hydraulic pressure in both direction.

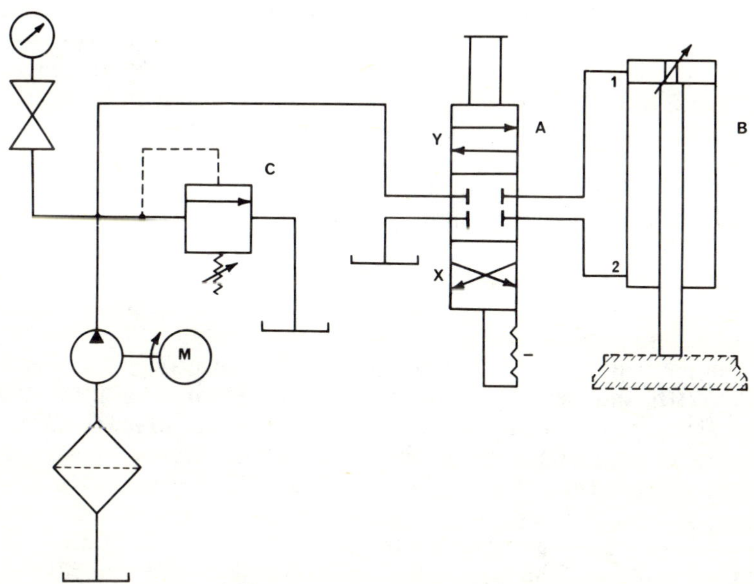

Fig. 3-3. Circuit with inching capability.

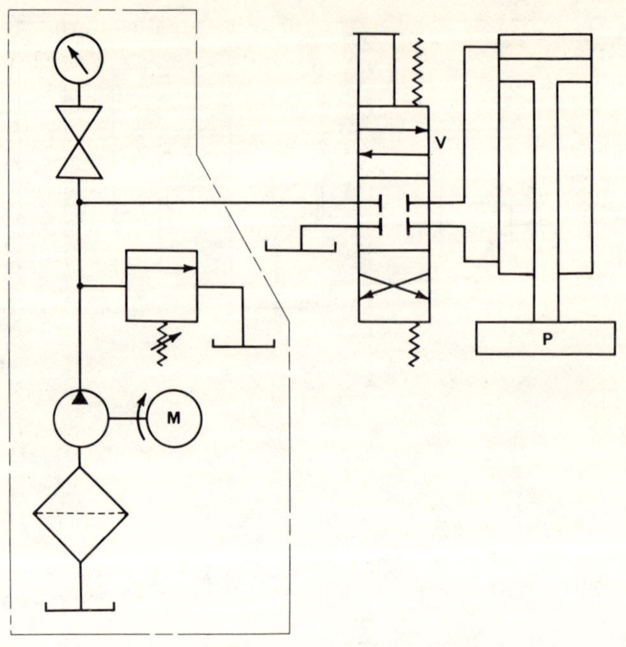

Fig. 3-4. Circuit with "deadman" control.

valve is depicted as a detented valve and will remain in whatever position the operator chooses. "Deadman" control can be achieved by employing a spring-centered valve as depicted in Fig. 3-4. Such a valve will effect hydraulic stability of the press platen at any time the manual valve operator is released.

The same inching characteristic can be achieved as shown in Fig. 3-5A. Three-position, closed-center, double-solenoid, spring-centered, directional control valve A offers a range of functional versatility that is difficult to achieve with manual valving. The closed-center condition provides the hydraulic stability necessary for inching, but it can be selectively engaged or disengaged at the discretion of the operator. The electrical diagram shown in Fig. 3-5B offers such selectivity.

With the selector switch (RUN-INCH) in the RUN mode, a straight-forward up-down characteristic is present. Momentary closing of the UP push button in line 1 sends a signal to the coil of relay CR1. Closure of the CR1 contacts in line 2 supplies a holding signal to the CR1 coil through the closed contacts of S1 and the normally closed contacts of the DOWN push button. The CR1 relay contacts of line 5 are also held closed, supplying a signal to solenoid X of directional control valve A. Hydraulic fluid is thereby supplied to the cap-end chamber of press cylinder C to create an upward movement and/or force.

Reversal of this action is accomplished by momentary closure of the DOWN push button in line 2. This closure completes a signal through its normally open contacts in line 3 to the

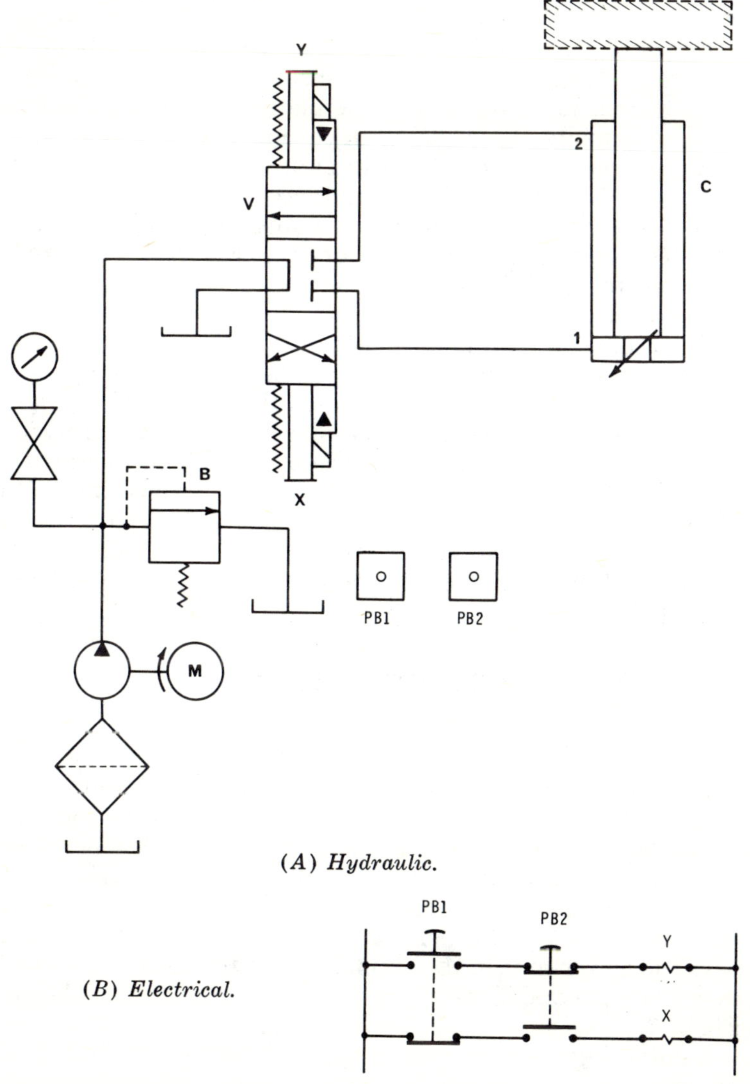

(*A*) *Hydraulic.*

(*B*) *Electrical.*

Fig. 3-5. Inching circuit using solenoid control.

coil of CR2. This closing action of DOWN push button simultaneously opens its normally closed contacts in line 2 and interrupts the holding signal to the coil of CR1. Closure of the CR2 contacts in line 4 supplies a holding signal to coil CR2 through the closed contacts of S1, the normally closed contacts of the UP push button, and normally closed limit switch LS. The normally open contacts of CR2 in line 6 are now held closed, supplying a signal to solenoid Y of directional control valve A. Hydraulic fluid is thereby supplied to the head-end chamber of the press cylinder to return its piston to the down position. Limit switch LS is opened by the platen in the down position. As the normally closed LS contacts in line 4 are opened, the holding signal to coil CR2 is interrupted, permitting CR2 to "drop out." Upon completion of the cycle, all relay coils and solenoids are de-energized, and the springs of valve A shift it to the center position. The point of press reversal is selected by the operator and initiated by momentary actuation of the proper push button.

To obtain the inching characteristic, the operator merely shifts the selector switch to the INCH mode. This opens the contacts in lines 2 and 4 and renders the relay holding signals inoperative. Press movement in either direction can be achieved only while one push button or the other is held operated.

The key to the inching function is the three-position feature of the directional control valve with its cylinder ports blocked in the center position. A variation of this valve configuration is discussed in Chapter 8 regarding the control of possible heat buildup.

COUNTERBALANCE

The circuits depicted in Fig. 3-4 and Fig. 3-5 offer a potentially serious problem that the designer should consider with respect to safety. If the weight of the vertical cylinder piston with its attached platen P is sufficient, a runaway characteristic might be encountered when the control valve is operated to move the cylinder piston downward. This characteristic could create a control problem as well as a safety hazard, since operation of valve V could permit the platen to drop rapidly even with the power unit in the off mode. This problem can be solved by the simple addition of counterbalance valve C as shown in Fig. 3-6.

Counterbalance valve C is normally closed until a preset pressure value is reached in the head-end cylinder chamber. This permits the setting of the counterbalance valve to match

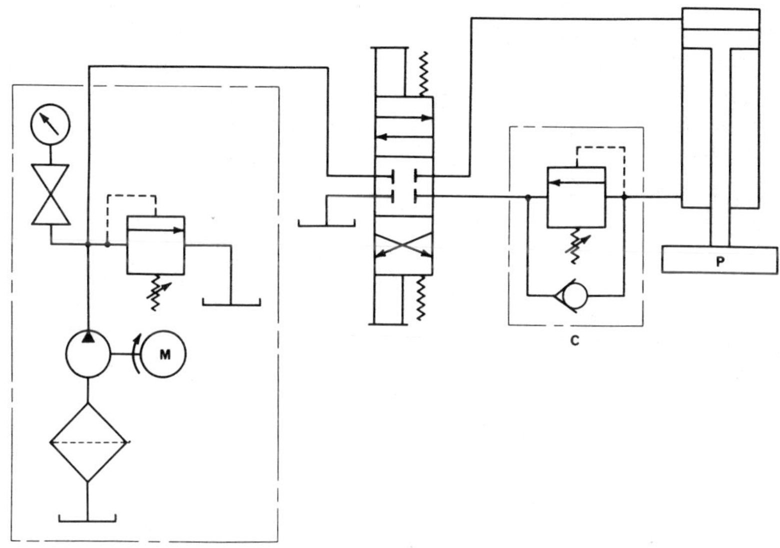

Fig. 3-6. Using a counterbalance valve to prevent runaway.

the weight-generated pressure and thus achieve an offsetting force value to cancel out the runaway characteristic. The check valve in the counterbalance package permits free flow to the head-end cylinder chamber, thus bypassing the counterbalance feature in that direction.

A second method of nullifying this runaway characteristic is depicted in Fig. 3-7. Here, pilot-operated check valve C in series with flow control valve F will produce similar results. The pilot-operated check valve prevents flow from the head-end cylinder chamber until pressure from the cap-end cylinder line pilots it open. This prevents the piston with attached platen from dropping without hydraulic power applied to the cap-end cylinder chamber. The purpose of the flow-control valve is to establish a maximum rate of piston travel and thus cancel the runaway characteristic. Without the flow control, or with its orifice setting too great, the runaway characteristic will result in a high-frequency open and close action of the pilot-operated check valve. The undesirable result will be a violent stuttering effect accompanied by severe hydraulic shock.

MANUAL START-AUTOMATIC RETURN

Semiautomatic operation of a hydraulic press frees the operator and eliminates the need to continuously monitor both the closing and opening strokes of the platen. The designer has three options available for initiating the return portion of the press cycle: (1) position initiation of return; (2) time initiation of return, and (3) pressure initiation of return.

Automatic return from a predetermined position or stroke length can be achieved in two primary ways: (1) mechanical pilot signal, or (2) electrical signal. Fig. 3-8 depicts a press circuit that employs four-way directional control valve A with a

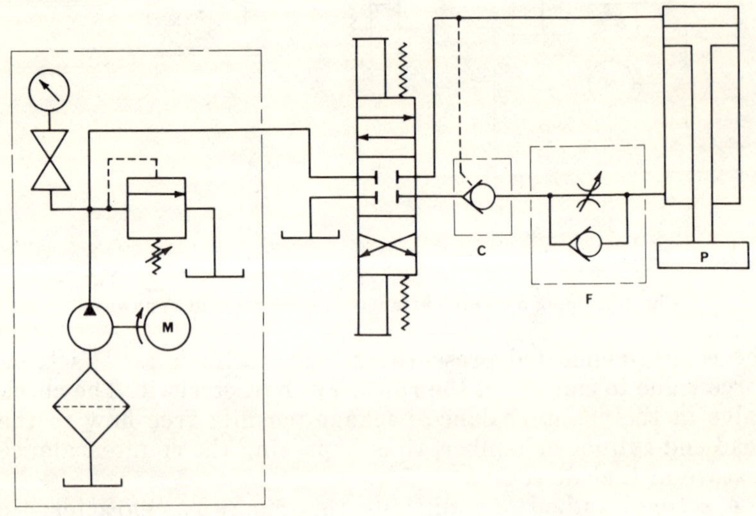

Fig. 3-7. Alternate method to offset runaway.

manual operator on one end and a hydraulic pilot operator on the other end. The cycle is started by manual actuation of four-way directional control valve A. This delivers fluid to the cap-end chamber of press cylinder C and powers the extension stroke of its piston. At the forward position of the piston stroke, cam E on the press platen actuates three-way, cam-operated, normally closed valve B to deliver a pilot fluid signal to pilot chamber X of four-way control valve A. As A shifts to its initial position, fluid is again directed to the head-end chamber of press cylinder C and powers the return stroke of the piston, thus completing the cycle. The length of stroke is determined by the point at which the cam-operated valve is actuated by the platen. The stroke length can be altered by adjusting the cam position on the platen.

Fig. 3-9A offers the same functional characteristics as does Fig. 3-8, except for the control valve operators and the signal medium. Valve A is a double-solenoid, two-position, four-way, directional control valve that is controlled by electrical signals. Limit switch LS1 is substituted for the pilot valve in Fig. 3-8, and the cycle is initiated by actuation of push button PB1. The hydraulic portion of the circuit is thereby simplified. Pump D has a variable volume adjustment feature that can be regulated to determine the speed of piston travel. The signal portion of the circuit is depicted in Fig. 3-9B1.

Momentary actuation of PB1 sends an electrical signal through the normally closed contacts of on-delay relay TR in

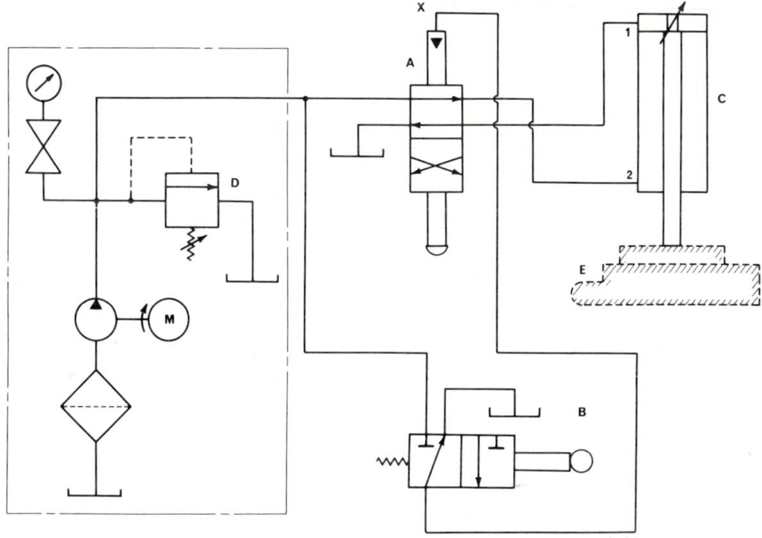

Fig. 3-8. Manual start, automatic return with hydraulic pilot operator.

line 1 to power solenoid X. This shifts valve A to initiate the down stroke of the press. At or near the end of the down stroke, limit switch LS1 is closed by contact with platen cam E, thus sending a signal to the coil of on-delay relay TR in line 2. When TR times out, a signal is completed through normally open TR contacts in line 3 to power solenoid Y of valve A. This returns valve A to its initial position, causing fluid to be delivered to the head-end cylinder chamber for return of the cylinder piston. The function of the on-delay relay TR is to delay the return signal and thereby ensure the completion of the full down stroke of the press. The time valve of the TR can be utilized to cause a dwell action at the closed position of the press if such a characteristic is desired. The combination of the normally closed TR contact in line 1 and the normally open TR contact in line 3 avoids the possibility of powering both valve solenoids simultaneously. If actuation of PB1 is continued beyond the time-out point of TR, a reciprocating action at the forward end of the stroke will result. The cycle will repeat at the instant cam E clears limit switch LS1.

An additional feature often required of a press circuit is that of a manually initiated, emergency return of the press platen. Such a provision can readily be added to the 3-9B1 circuit as shown in Fig. 3-9B2. The normally open switch contacts in line 4 permit an optional method of signaling return

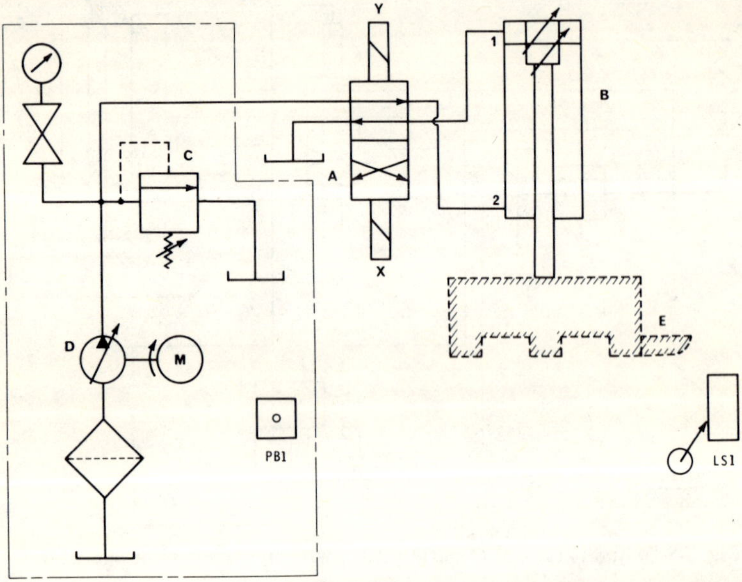

(A) Hydraulic.

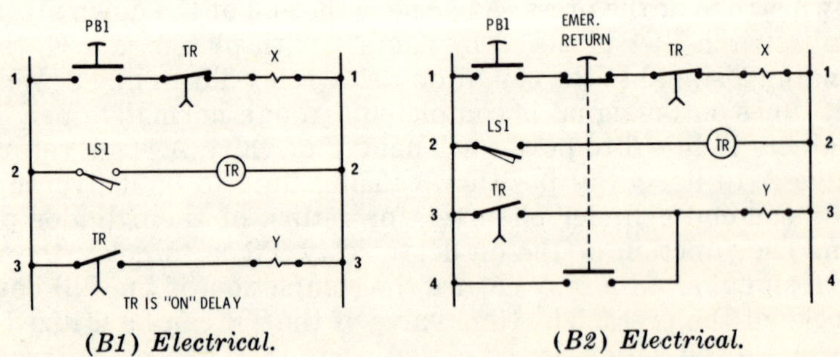

(B1) Electrical. *(B2) Electrical.*

Fig. 3-9. Manual start, automatic return with limit switch.

solenoid Y, should it become necessary to reverse the press stroke prematurely. The normally closed switch contacts in line 1 will avoid the possibility of powering both valve solenoids simultaneously.

An alternate method of achieving a mechanically actuated pilot signal for automatically initiating the return stroke is illustrated in Fig. 3-10. Valve A is a four-way, hydraulic, two-position directional control valve with an air pilot operator

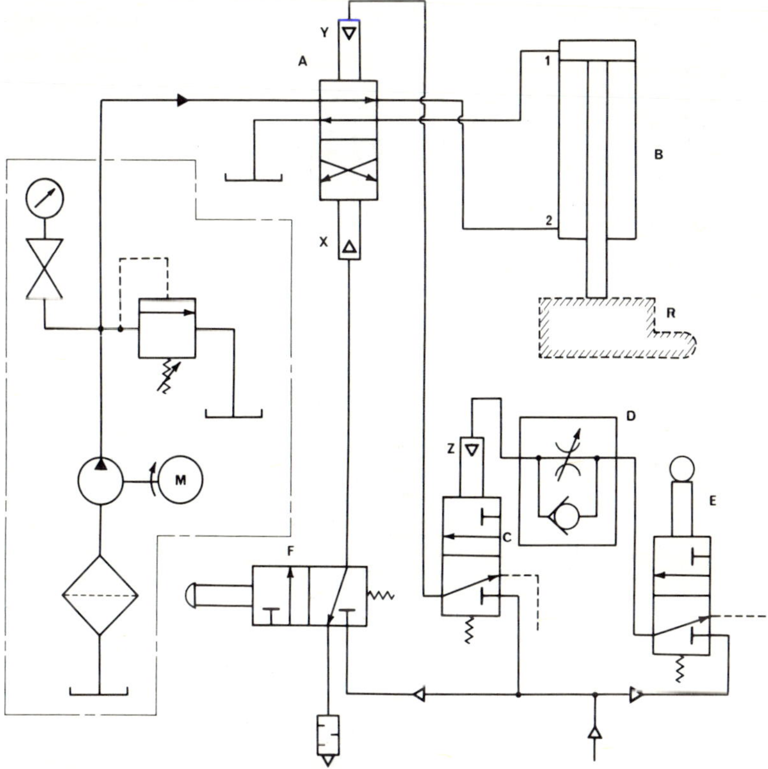

Fig. 3-10. Air-pilot operation for manual start and automatic return.

on each end. All control signals are air, while the power medium is hydraulic. Operation of hand-operated valve F supplies a signal to pilot chamber X, which shifts valve A to power the extension stroke of the press platen. As the press platen reaches the end of its forward stroke, cam R actuates air-pilot valve E and starts an air signal to pilot chamber Z of valve C. This signal is delayed by flow-control valve D to ensure the completion of the full down stroke of the press. When the delayed shift of valve C is accomplished, an air signal is supplied to pilot chamber Y of valve A to return it to its initial position. Hydraulic fluid is thus supplied to the head-end chamber of press cylinder B to return the piston and platen to the *up* position and complete the cycle.

The emergency return feature can be added to this circuit as illustrated in Fig. 3-11. Shuttle valve S permits a signal to pilot chamber Y from either of two sources. For normal reversal of the press stroke, the pilot signal is supplied by valve C as in the Fig. 3-10 circuit. If the need develops to reverse the press

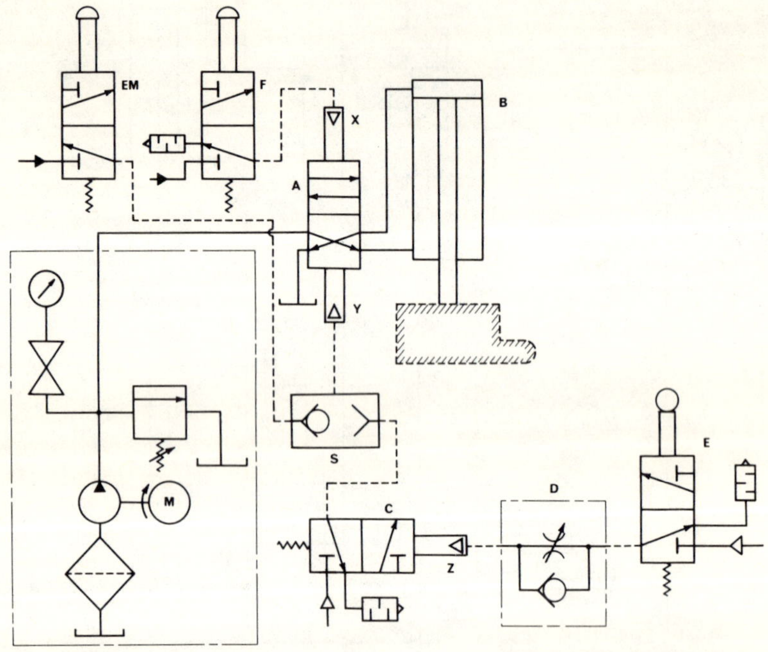

Fig. 3-11. Circuit demonstrating air-actuated emergency return.

stroke prematurely, the reversing signal can be supplied through hand-operated, emergency-return valve EM. Regardless of which pilot valve initiates the signal, the shifting of the shuttle valve isolates the other from the circuit.

It is sometimes desirable to provide for automatic return of a press platen immediately upon reaching a predetermined force value. This is particularly true if a wide range of dimensions of the load is encountered. One example would be press fitting assemblies that require a maximum interference limit. The simplest method of accomplishing this characteristic is depicted in Fig. 3-12A and Fig. 3-12B. Valve A is a four-way, two-position, double-solenoid, directional control valve. Pressure switch C is installed in the cap-end cylinder line and set to close at the hydraulic pressure that creates the maximum desired force of the cylinder ram. Push-button switch PB1 has a set of normally open contacts and a set of normally closed contacts. Initiation of the press closing is accomplished by momentary closing of the switch contacts in line 1 of Fig. 3-12B to supply an electrical signal to solenoid X of valve A. Auto-

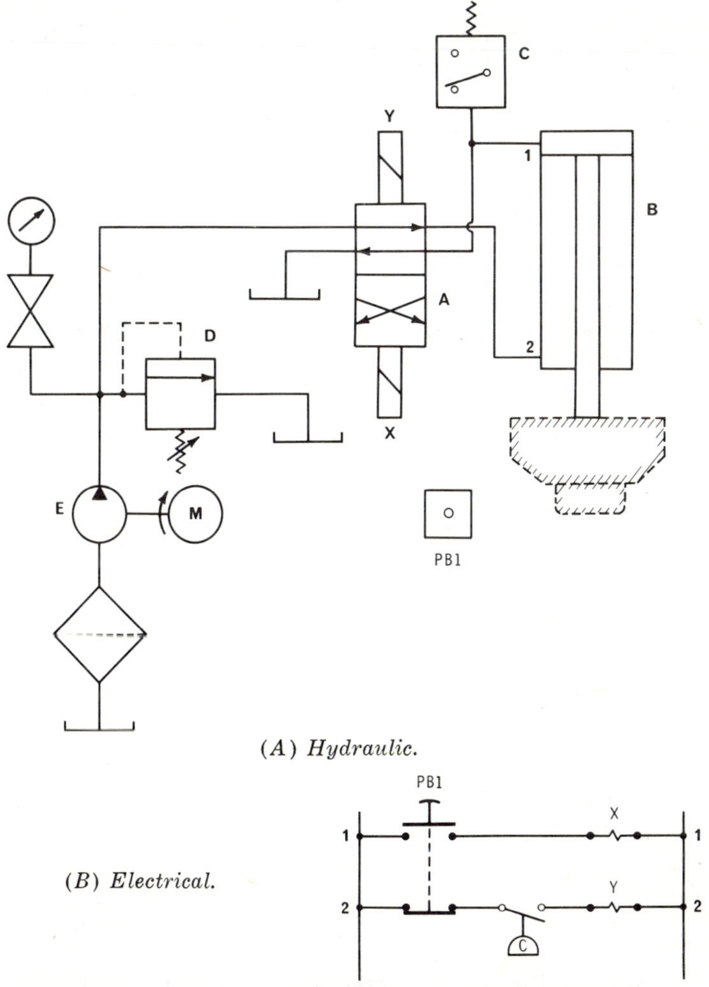

(*A*) *Hydraulic.*

(*B*) *Electrical.*

Fig. 3-12. Automatic return using pressure-operated switch.

matic reversal is achieved by pressure closure of pressure switch C, thus supplying an electrical signal through the normally closed contacts of PB1 in line 2 to solenoid Y of valve A. If PB1 is maintained in an operated position, pressure switch C cannot supply the signal and will thereby be overridden. This not only permits the override feature, but prevents energization of both solenoids simultaneously.

TWO-HAND CONTROL

Safety of the operator is often a critical factor in the functioning of a hydraulic press, particularly if exposed pinch points are left unguarded. One practical approach to the problem is to design the control circuit to require the use of both hands to initiate and sustain the closing movement of the press platen. Employing the hydraulic circuit depicted in Fig. 3-5A, two-hand control can be achieved with the electrical circuit depicted in Fig. 3-13.

In the idle state, normally open limit switch LS2 in line 2 is held open by the press platen in the down position. The coil of relay CR in line 4 is energized by the signal through the normally closed contacts of PB1, PB2, and TR. Simultaneous operation of PB1 and PB2 will complete a signal through the normally open contacts of PB1 in line 6, the normally open

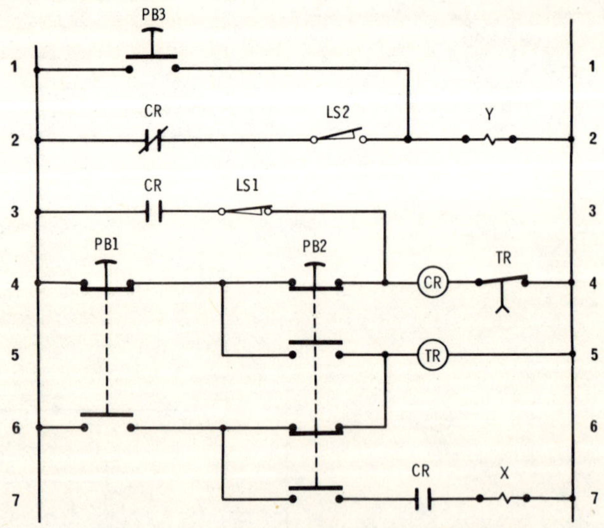

Fig. 3-13. Electrical circuit for two-hand control (see Fig. 3-5 for hydraulic).

contacts of PB2 in line 7, and the normally open, held closed contacts of CR in line 7 to solenoid X. With PB1 and PB2 actuation maintained, valve A remains in the shifted position supplying fluid to the cap-end cylinder chamber for the extension stroke of the press ram. Relay coil CR in line 4 is held energized by the signal supplied through the holding CR contacts in series with normally closed LS1 in line 3, until the press platen opens LS1 in the up position of the press ram. Opening LS1 interrupts the holding signal to coil CR in line 4, which drops out and breaks the signal to solenoid X in line 7. When CR drops out, a signal is supplied to solenoid Y in line 2 through the normally closed contacts of CR and normally closed LS2 to shift valve A to the press return position. When the press platen is fully returned, LS2 is held open, thus breaking the signal to solenoid Y and completing the cycle.

Both PB1 and PB2 must now be released before another cycle can be initiated. The press can be inched by momentary actuation of PB1 and PB2 simultaneously. If PB1 is "tied down," TR in line 5 receives a signal through line 6, causing it to time out. If PB2 is tied down," TR in line 5 receives a signal through PB1 in line 4 and PB2 in line 5, causing it to time out. In timing out, normally closed TR contacts in line 4 open and cause CR to drop out, rendering the circuit inoperable until PB1 and PB2 are both released.

Completion of the full cycle requires the maintained actuation of both PB1 and PB2. If both switches are released during the closing stroke of the press, the platen will stop and hold position until they are reactuated. The press will then resume its closing stroke. If both PB1 and PB2 are released during the opening stroke of the press, the press will stop and remain in a partially open position until PB3 is actuated to supply an alternate signal to valve solenoid Y. For safety reasons, the designer might want to locate PB3 in a remote position away from the press. Should only PB1 or PB2 be released during the opening stroke of the press, TR will time out, causing CR to drop out. This drop-out action will supply a signal through the normally closed CR contacts and LS2 in line 2 to solenoid Y, thus initiating the opening stroke of the press. This characteristic can be utilized for the return of the press platen prior to the full completion of its closing stroke. If the other PB is released, however, the completion of the opening stroke can be accomplished only by actuation of PB3.

The foregoing two-hand control circuit can readily expanded to a two-operator, four-hand configuration merely by pyramiding the PB stations.

CHAPTER 4

Circuits for Controlling Rate of Movement

The rate of movement of a hydraulic force component is determined by the volume of fluid delivered to or allowed to pass from its chamber (s) in a unit of time. Whether the component is a linear unit or a rotary unit, the rate of movement will vary in direct proportion to any induced change in the volumetric flow rate of the fluid powering the component. The problem of controlling rate of movement is simply the problem of controlling fluid flow rate either to or from the component being controlled.

In a broad sense, varying the relationship of cylinder bore and pump displacement can effect a change in the rate of travel of the cylinder piston. For example, a given size of fixed-displacement pump will create a rate of piston travel four times faster in a two-inch bore cylinder than it will in a four-inch bore cylinder. The piston speed ratio is inversely proportional to the ratio of the volumes of the two cylinders. The piston of a single-rod end cylinder will move faster on its retraction stroke than on its extension stroke, since the cap-end chamber is of a greater volume than the head-end chamber due to the volume displaced by the rod. While such factors have an important effect on rates of movement, it would be quite impractical to attempt to determine speeds by altering the displacements of components alone.

VARIABLE-VOLUME PUMP

Some applications do lend themselves to the technique of varying the effective displacement of components, and when this technique is applicable, it can be quite effective. Fig. 4-1 depicts a circuit that employs a hydraulic pump driving a hydraulic motor with no intermediate valving. Pump P is a variable-displacement, manual-control unit with a volumetric displacement that is subject to change by a hand-wheel adjustment. Decreasing the pump displacement will reduce the volume of fluid being delivered to hydraulic motor HM, and thereby reduce the rotational speed of the motor shaft. If the pump is capable of being adjusted in each direction from the zero displacement setting, both speed and direction of rotation can be controlled by the hydraulic pump stroke adjustment. Such a circuit falls into the category of hydrostatic transmissions and is expanded upon in Chapter 10.

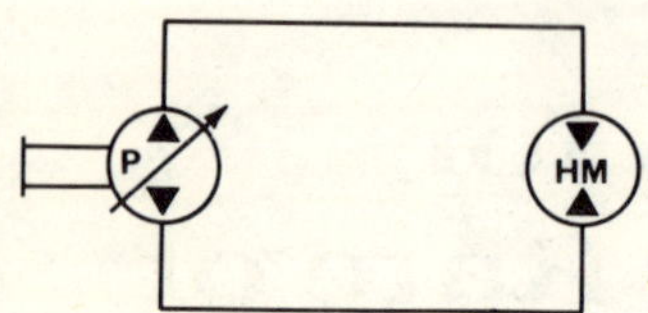

Fig. 4-1. Variable-displacement pump driving a hydraulic motor.

Fig. 4-2 depicts a variable-volume, manual-control hydraulic pump as the fluid source driving a hydraulic cylinder. When directional control valve A is shifted, the speed of piston travel of hydraulic cylinder C is determined by the displacement setting of pump P. For any given displacement setting, the pump functions as a fixed-displacement pump; therefore, relief valve D is required to protect the system against overpressure. The functional principle of the relief valve is discussed in Chapter 5.

FLOW-CONTROL VALVES

The most common method of controlling rates of movement in a hydraulic system involves the use of adjustable orifices. While many hydraulic components are equipped with built-in orifices to accomplish specific functions, the flow-control valve is a packaged orifice designed to offer the maximum in control and ease of installation in a hydraulic circuit. The flow-control package contains a minimum of two features: (1) an adjustable orifice or needle valve and (2) a check valve for uninhibited reverse flow. Additional optional features are automatic compensation for pressure and temperature variations.

PRESSURE-COMPENSATED FLOW CONTROL

The proper selection of a flow-control valve requires an understanding of compensation versus noncompensation. A given orifice setting will pass a varying amount of fluid, depending on the pressure difference across the orifice. The greater that pres-

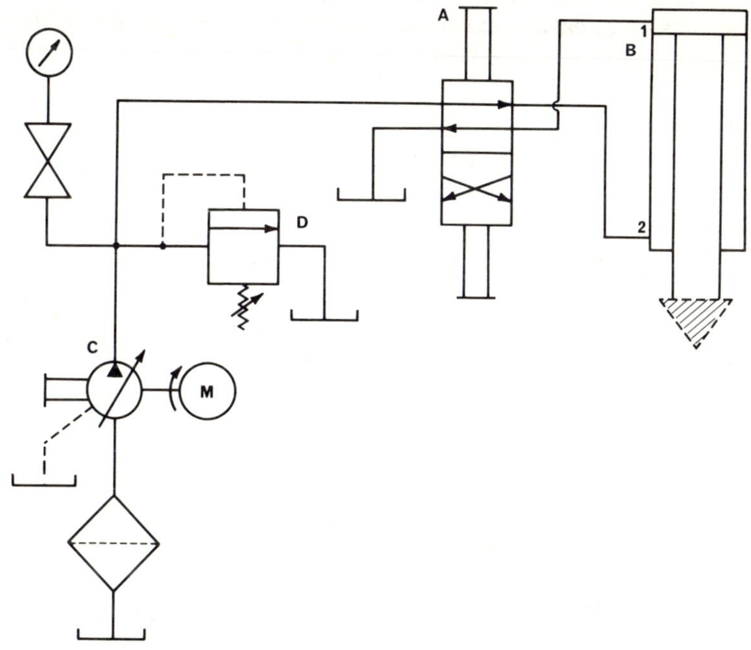

Fig. 4-2. Flow control with a variable-volume pump.

sure difference, the greater will be the fluid flow rate. A pressure-compensated orifice incorporates a pressure-sensitive, self-regulating provision that reduces the orifice size as the pressure differential across it increases. The effect of this compensation is to maintain a comparatively constant fluid flow rate, regardless of changes in the pressure differential. While this feature is highly desirable in maintaining a constant rate of movement against a variable load resistance, not all flow-control applications require such constancy, nor is it necessarily desirable. The alert circuit architect will use descretion by first analyzing carefully the performance desired.

In Fig. 4-3, a pressure-compensated, flow-control valve is employed to establish the rate of movement as the cylinder piston extends. Flow-control valve C is placed between four-way, directional control valve A and the cap-end cylinder port to control the piston speed in one direction only. During the return stroke of the cylinder piston, the check valve of flow-control valve C permits unrestricted flow from the cap-end chamber, through control valve A, and to the reservoir. The location and metering direction of flow control C identifies this as a "meter-in" or input flow-control technique. The pressure-compensating feature causes a constant extension speed of the

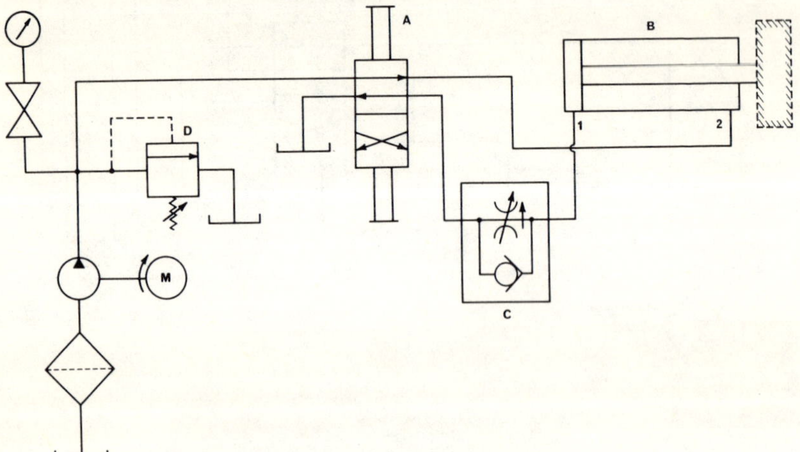

Fig. 4-3. A circuit illustrating input flow control.

cylinder piston, regardless of the possible inconstancy of the load resistance. The extension speed might be anything less than full pump displacement, while the return stroke speed is determined by, and equal to, the full pump displacement.

Fig. 4-4 incorporates a noncompensated flow-control valve that is located to control the rate of fluid flow *from* the head-end chamber during the extension stroke of the cylinder piston. Directional control valve A is a four-way, open-center, three-position, spring-centered configuration that interconnects all ports to reservoir in its neutral position. When control valve A is shifted by the X actuator, fluid is delivered to the cap-end chamber of cylinder B, causing the piston to extend. The rate of extension is determined by the adjustment of the orifice of flow-control valve C, limiting the fluid flow rate from ahead of the piston back to the reservoir. This technique is identified as "meter-out," or "output flow control."

To retract the piston of cylinder B, control valve A is shifted by the Y actuator, and fluid is directed through the check valve of flow-control valve C to the head-end cylinder chamber. The control orifice of flow-control valve C is nonfunctional during the cylinder retraction stroke. A typical application for this circuit might be a clamping operation, a push operation having a constant load resistance, or a simple "over-center" movement requiring simple control of an overrunning load. Deactivation of both valve operators will remove pressure supply from both cylinder chambers, but it will permit piston travel from either its own load momentum or the application of some outside force. If a "locked" piston condition is desired, the directional control valve should incorporate the closed-center configuration depicted in Fig. 4-6.

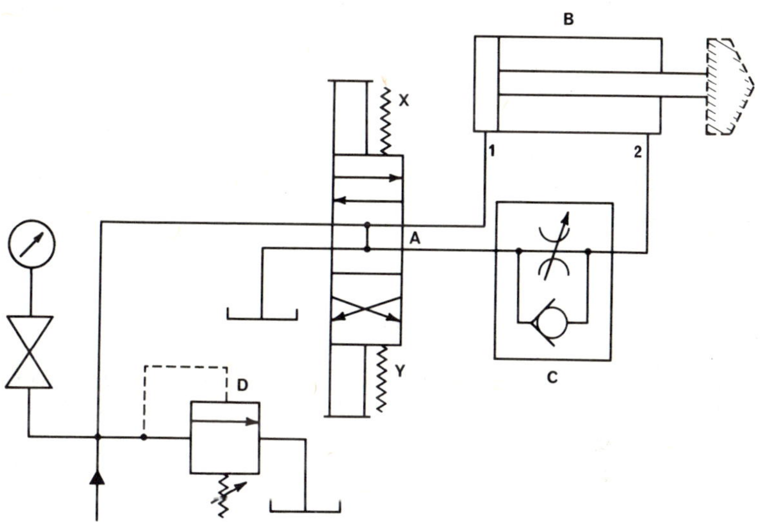

Fig. 4-4. A circuit illustrating output flow control.

The Fig. 4-5 circuit employs pressure-compensated, flow-control valve B to divert a portion of the extension stroke supply volume away from the cylinder C cap-end chamber and directly to the reservoir. This diversion technique is identified as "bleedoff" flow control and affects the rate of piston travel by taking from the supply volume powering the action. When two-position, four-way, directional control valve A is shifted, fluid supply is directed to the cap-end chamber of cylinder C to extend its piston. This supply fluid volume is reduced by the amount the orifice of flow control B permits to return back to the reservoir. When the directional control valve operator is released, fluid is directed to the head-end chamber of cylinder C to power the retraction stroke at a rate determined by the fluid supply volume. During this retraction stroke, the flow-control valve is nonfunctional. No check valve is necessary, as all flow through valve B is in one direction. For best results and most consistent regulation, the pressure-compensating feature is recommended. Without it, the rate of piston travel would vary inversely to the amount of load resistance. This technique would not be effective in controlling the "over-center" type of load referred to in the earlier paragraph.

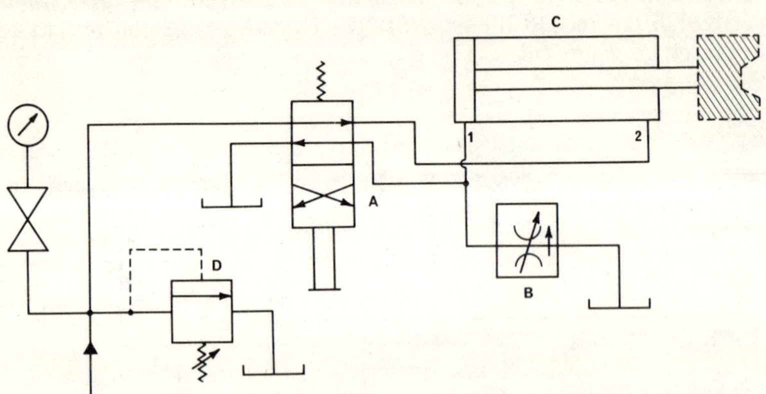

Fig. 4-5. A circuit illustrating bleed-off flow control.

Bidirectional speed control can be achieved simply by employing two flow-control valves as depicted in Fig. 4-6. Meter-out flow control is illustrated, with the orifice of flow-control valve D regulating the extension speed and the orifice of flow-control valve C regulating the retraction speed. The check valve options shown permit free flow of fluid supply to the cylinder chambers, thus rendering the flow-control orifices nonfunctional except when carrying fluid flow from the cylinder chamber to the reservoir. The rate of travel in either direction can be established without affecting speed in the opposite direction.

FAST APPROACH-TWO SPEED

Speed changes during a hydraulically powered movement can be accomplished by the use of control orifices that can be activated on command. A typical application might be a rapid approach to the work position, and a slow feed rate through the work station. Fig. 4-7 depicts the use of a normally open (normally inactive) flow-control valve that is activated by cam action of the member being moved. When four-way, directional control valve A is shifted, fluid is directed to the cap-end chamber of cylinder B, powering its extension stroke. Fluid passes from the head-end chamber through the open porting of cam-operated, flow-control valve C to reservoir. The speed of the extension stroke is determined by the rate of supply volume until cam action closes the normally open porting of flow-control valve C. When C is closed, the rate of fluid flow from the head-end cylinder chamber is determined by the adjustable orifice of the flow-control valve. Reversal of directional control

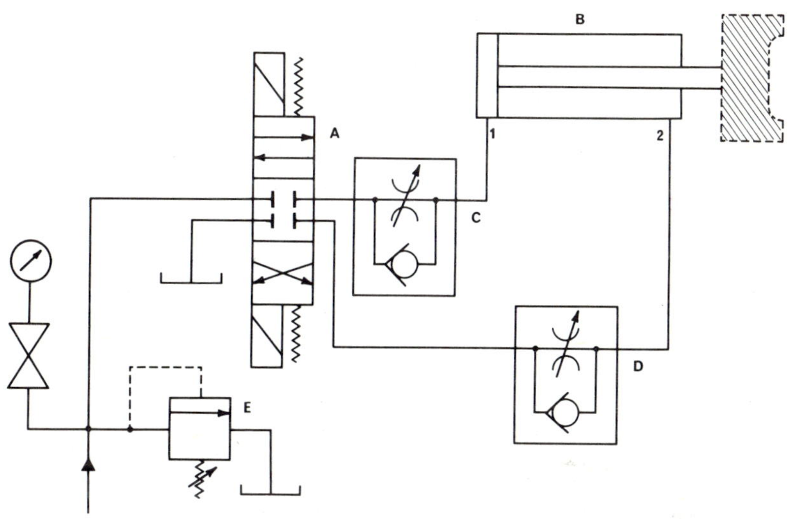

Fig. 4-6. Bidirection flow control using flow-control valves

valve A supplies fluid through the check valve of flow-control valve C to the head-end chamber of cylinder B to power the unrestricted return of its piston. The net results are rapid approach, controlled feed, and rapid return.

The circuit depicted in Fig. 4-8A is identical in function to the Fig. 4-7 circuit, except that each movement is initiated by an electrical signal. The extension stroke is initiated by momentary closure of PB1, and the retraction stroke is initiated by momentary closure of PB2. By connecting a normally open contact of each push button in series with a normally closed contact of the other push button (lines 1 and 2 in Fig. 4-8B), they will cancel each other out. This series arrangement eliminates the possibility of energizing both valve A solenoids simultaneously, resulting in solenoid coil failure. The controlled feed portion of the extension stroke is initiated by closure of limit switch LS1 in line 3, sending an electrical signal to solenoid Z to activate flow-control valve C. The check valve of flow-control valve C renders the valve C orifice nonfunctional during the entire retraction stroke, even though solenoid Z remains energized until the cylinder B cam leaves LS1.

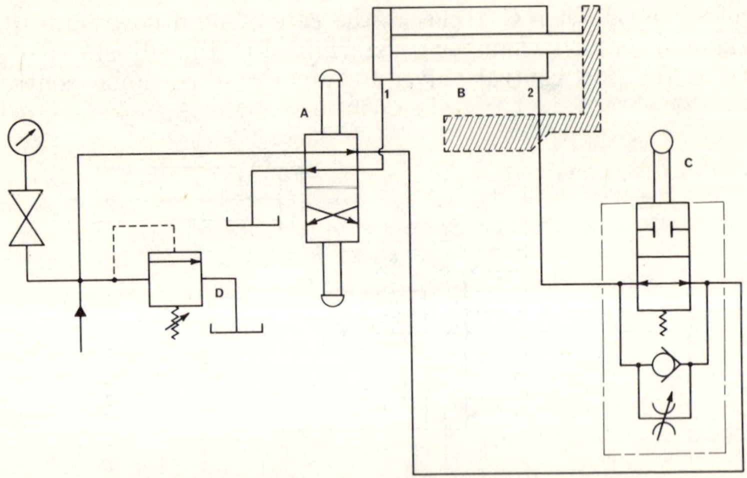

Fig. 4-7. Circuit using a cam-operated, flow-control valve.

DECELERATION

When a cylinder is permitted to move through its stroke or strokes at a relatively high speed, contact of the piston with the cylinder cover can result in severe mechanical as well as hydraulic shock. Where full piston stroking is used, a satisfactory dampening effect can often be obtained by the use of cylinder cushions. A cylinder cushion is a built-in orificing device that restricts the fluid flow from the cylinder chamber as the piston nears the end of its stroke. Standard cushions are usually effective through the final one-half to one inch of piston travel, and the orifice is usually adjustable. If the average cushion length is not adequate, as in the case of long strokes, high speeds, and high inertial loading, longer than standard cushions are available. Such special cushions may be of any reasonable length, but usually do not exceed six to eight inches in length. These cushions are designed to gradually decrease the exhaust orifice as the piston approaches the cylinder cover, thus providing gradual deceleration. Fig. 4-9 depicts a simple circuit that incorporates extra-long adjustable cushions for each stroke direction of cylinder B.

A varied approach to the deceleration problem could be to employ a low-angle cam to actuate flow-control valve C in Fig. 4-7. Such a cam operation would gradually close off the exhaust flow path to reservoir, and the effectiveness can readily be varied by altering the cam angle and length.

Deceleration for each stroke direction is illustrated in Fig. 4-10A. Cylinder B powers machine table T and is controlled by two-position, double-solenoid, four-way, directional control valve A. In its idle state, cylinder B is in the retracted position as shown, and LS1 is in the actuated position. The automatic reciprocating cycle is started by momentary closure of PB1 in line 1 (Fig. 4-10A), sending a signal to the control relay coil

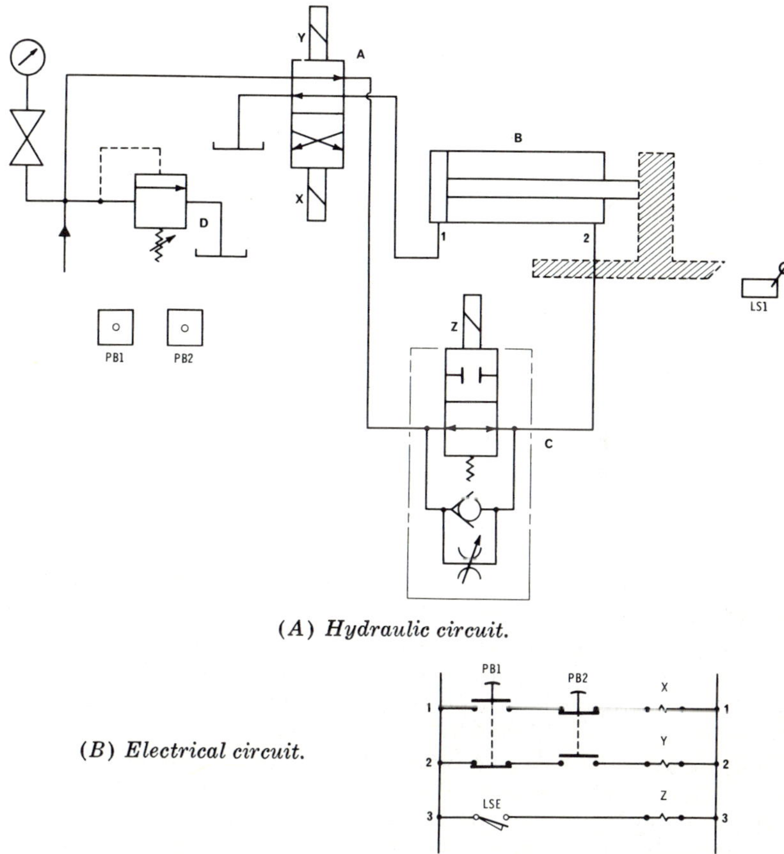

(*A*) *Hydraulic circuit.*

(*B*) *Electrical circuit.*

Fig. 4-8. Circuit using a solenoid-operated, flow-control valve.

CR. A holding signal is effected through the CR contacts in line 2 in series with the normally closed PB2 contacts. The resulting closure of the CR relay contacts in line 4 completes a signal through the normally open (held closed) contacts of limit switch LS1 to solenoid Y of valve A. As valve A shifts, fluid is directed to the cap-end chamber of cylinder B, powering its extension stroke. The LS1 contacts in line 4 reopen as soon as the machine table begins it extension stroke. Rapid travel is in effect until cam action of the machine table activates cam-operated, flow-control valve D as it nears the end of its extension stroke. The gradually induced meter-out action of flow-

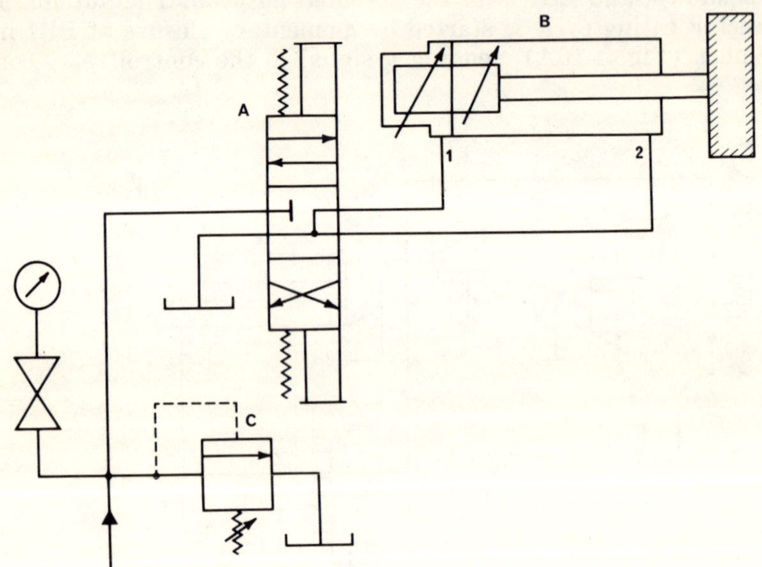

Fig. 4-9. Diagram illustrating the use of adjustable cushions for deceleration.

control valve D effects a gradual deceleration of the table traverse. At the end of the forward stroke, LS2 in line 2 is momentarily closed, completing a signal to solenoid X of control valve A. Valve A is thus shifted to power the return stroke of the table. Near the end of the return stroke, cam-operated, flow-control valve C performs the same meter-out, deceleration function as did valve D for the extension stroke. Upon again making contact with LS1, the cycle is reinitiated and will continue to automatically reciprocate until PB2 is actuated to "drop out" the CR relay coil. Since only the Y solenoid signal is affected by the CR contact closure, the table will always

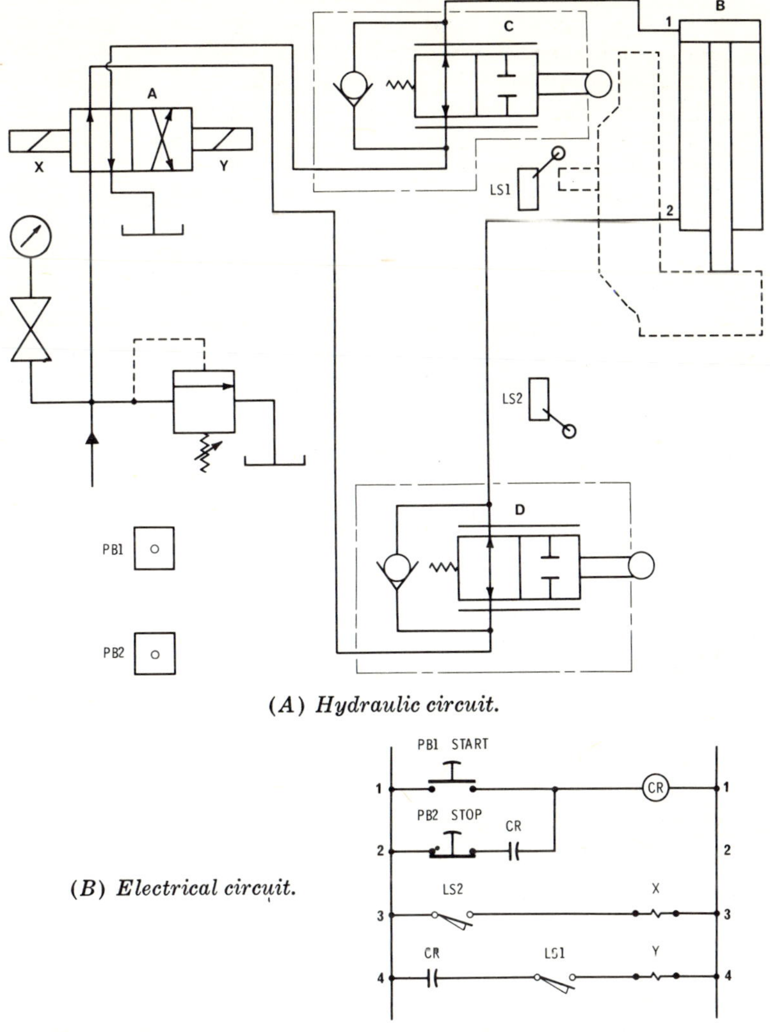

(A) Hydraulic circuit.

(B) Electrical circuit.

Fig. 4-10. Circuit illustrating deceleration through the use of valves.

come to rest in the cylinder retracted position until a new reciprocating cycle is initiated by momentary closure of PB1. For a more sophisticated approach to the problem of deceleration, see Chapter 11.

CHAPTER 5

Circuits for Establishing Pressures and Forces

The forces obtained in a hydraulic circuit are the product of the hydraulic pressure and the effective movable area against which the fluid is pushed. A hydraulic pump merely displaces a volume of fluid. Pressure is a measure of resistance to the flow of that fluid. When the designer attempts to limit the pressure generated in a hydraulic system, he is merely establishing a maximum value of the flow resistance.

The size of fluid conduit lines employed in a circuit can create a wide range of flow resistance values, but these values are not usable or controllable when established by such a non-adjustable limit to flow. The wise designer will select optimum line sizes that effect a minimal flow resistance, reserving the pressure control parameter for the controllable componentry that has been specifically designed to facilitate ease of selectivity. He is strongly urged to utilize conduit sizes that permit flow rates not to exceed 15 feet per second in pressure lines and 5 feet per second in return lines. If extremely long conduit lines must be employed, these flow values should be reduced accordingly by increasing the line diameters.

ESTABLISHING PRESSURE

The most fundamental component for limiting pressure in a hydraulic system is the relief valve. It remains as a normally closed, nonfunctioning unit until the flow resistance within the system reaches a preset limit. Since a hydraulic system (unlike a pneumatic system) is a "closed" system, one of the functions of the relief valve is to provide (with its attached plumbing) an alternate flow path to the reservoir. It does so only when the normal flow path within the circuit offers a magnitude of flow resistance greater than the pressure value established by the relief-valve spring. In other words, the relief valve opens an alternate flow path to the reservoir at any time the normal working conduit lines are unable to accept all of the fluid being displaced by the pump at a preset flow resistance or pressure

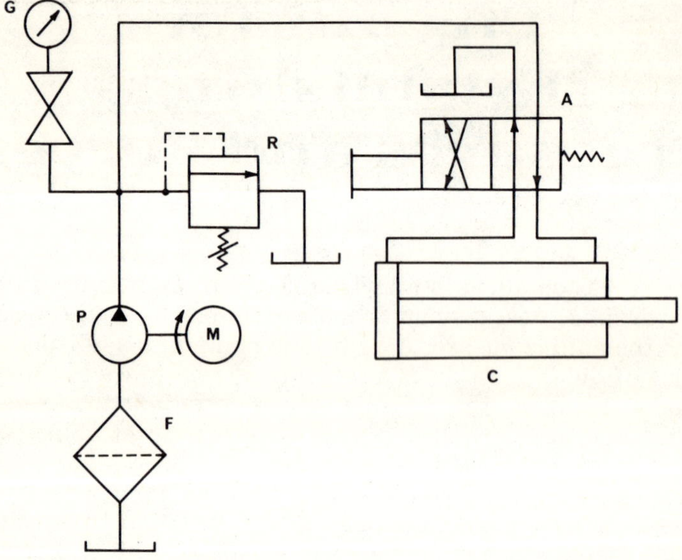

Fig. 5-1. Hydraulic circuit using pressure-relief valve to establish pressure.

value. This value can be varied simply by varying the force value of the spring. Whenever the normal fluid channels within the working portion of the circuit are again able to carry all the fluid being displaced by the pump, the relief valve returns to its normally closed condition.

In Fig. 5-1, relief valve R performs the function described in the foregoing paragraph. Pump P is a fixed-displacement, rotary unit revolving at a constant rpm and driven by motor M. A specific volume of fluid is moved per unit of time from the reservoir, through filter F, and into the conduit lines of the system. When valve A is shifted, the fluid is allowed to flow through its channels to the cap-end cylinder chamber at a rate that coincides with the displacement rate of the pump. The resistance to flow is the sum of the fluid friction within the conduit lines, the mechanical friction of the moving cylinder parts, and the load resistance impeding the cylinder piston travel. When the load resistance rises excessively, or when the cylinder piston reaches the limit of its travel (in either direction), the fluid being displaced by the pump has no place to go. The resistance to flow has suddenly become theoretically absolute. At this moment, the pressure rise in the system pilots relief valve R to its open position, thus providing an alternate path for the pumped fluid. This alternate path merely returns the "unused" fluid to reservoir. At this time, the pressure value in the cylinder and the effective conduit lines is determined by the opening pressure value established by the relief-valve spring setting. The size of the opening that the relief valve presents, to the fluid delivery to it, automatically regulates itself to maintain the pressure value desired. The moment a flow path that offers flow resistance of a lower pressure value is made available (such as the reversal of the control valve), the relief valve immediately closes.

The cylinder depicted in Fig. 5-1 will exert a greater force on its extension stroke than on its retraction stroke because of the difference in the effective piston areas. The larger the

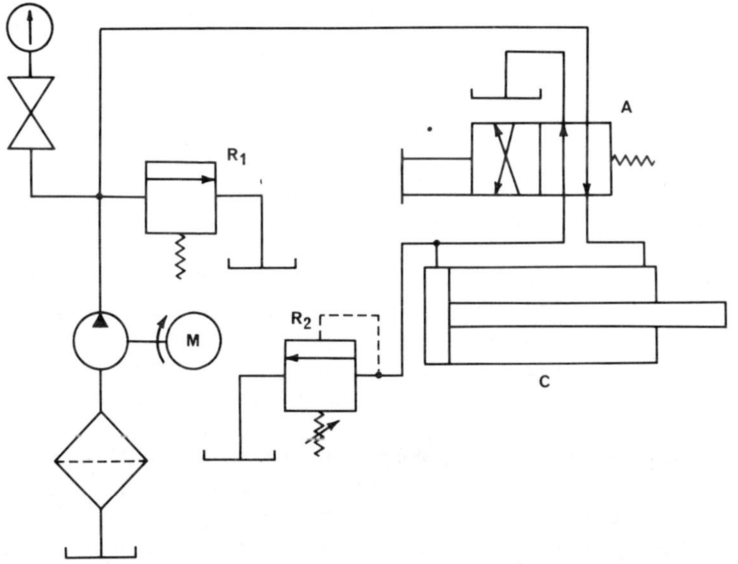

Fig. 5-2. Hydraulic circuit using two relief valves to establish pressure in both directions.

piston rod, the greater will be the difference between the extension and retraction forces. If the work requirement demands an equal force in each direction, the Fig. 5-2 circuit suggests a method of accomplishing this equal force through unequal pressure. Since pressure × area in one direction must equal pressure × area in the other direction, the pressure against the larger or cap-end area must be a value lower than that presented to the head-end area. Relief valve R2 is utilized to establish this lower value only in the supply to the cap-end chamber of cylinder C. With directional control valve A in the spring-offset position shown, relief valve R1 establishes the higher pressure value for the head-end chamber, and relief

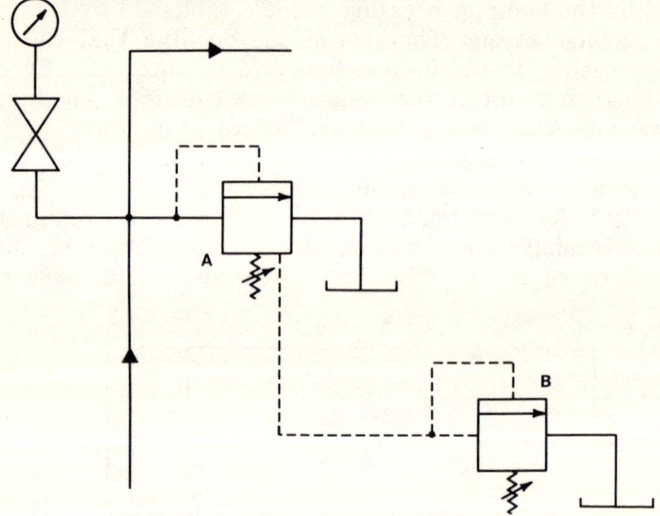

Fig. 5-3. Using a remote pilot, pressure-relief valve to set pressure.

valve R2 is nonfunctional. This is a satisfactory method of meeting the equal force requirement, unless the maximum pressure established by relief valve R1 is required at some other point in the system. See Fig. 5-5 for an alternate method of accomplishing such a force-balancing result in a multiple-function circuit.

The pressure value of relief valve R depicted in Fig. 5-1 is established by adjusting the spring value of its pilot cap. If a system requires frequent changes of this pressure value, the control adjustment can be made more accessible by utilizing a remote pilot unit. The remote pilot unit depicted in Fig. 5-3 is a relatively small unit required to carry only the pilot fluid volume of the control section of the main relief valve. Main relief valve A may be of considerable size, requiring large conduit lines and can be located at the pump-reservoir package. Remote unit B can be located at the operator's station for ready accessibility and convenience, as much smaller conduit lines are required to make its hookup. The setting of valve A establishes a maximum pressure value, and the adjustment of valve B can establish any pressure value below the setting of valve A, but it cannot effect any pressure in excess of that already established as a maximum by valve A.

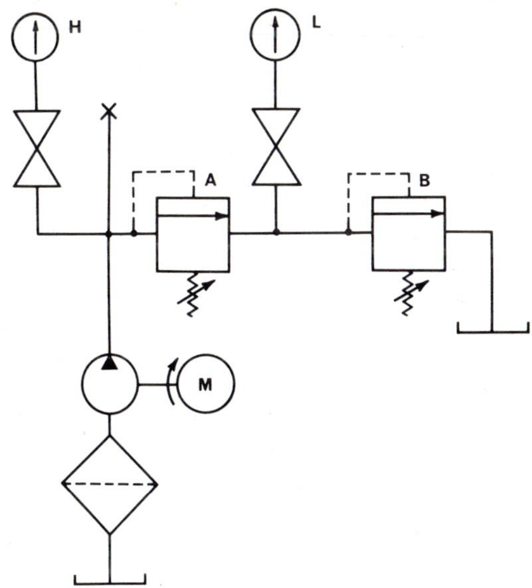

Fig. 5-4. Using two relief valves in series.

The relief valve primarily establishes a "differential" pressure across its piston. Therefore, any flow resistance in its discharge conduit will be reflected directly to its primary or inlet side. Fig. 5-4 illustrates two relief valves in a series or tandem arrangement. The discharge or tank port of relief valve A is connected to the inlet or pressure port of relief valve B. If A alone is given a differential setting of 500 psi, it will establish that as a maximum reading at gage H. If B is then given a differential setting of 300 psi, that pressure value will be reflected in the indication of gage L. The initial 500-psi value at gage H will immediately increase to 800 psi by virtue of the differential feature. At this point, a two-pressure system will have been created with some interesting characteristics. Adjustment of the differential setting of valve A alone will alter the pressure value at gage H only. However, by adjusting the pressure setting of relief valve B alone, both gage readings will be affected equally.

The advisability of using this Fig. 5-4 circuit to create a two-pressure system may be open to question in most cases, however, since the lower pressure can be maintained only by first reaching the higher pressure value. Therefore, all fluid is being delivered by the pump against the higher pressure or flow resistance. The arrangement might prove satisfactory if the extra input horsepower and greater heat generation is not a concern.

The pressure-reducing valve has been designed for the specific purpose of effecting secondary pressure values for a portion or portions of a hydraulic circuit. It offers a normally open flow path to the system, and the open condition is maintained directly or indirectly by a spring. The spring usually has an adjustable value. Whenever the downstream pressure reaches a sufficient value to overcome the spring force, that pressure is used to close the flow path by a pilot action. If the volume demand downstream has been satisfied, the valve closes completely and permits the passage of no more fluid. If an additional flow volume is required, the piloting action versus spring value modulate each other to permit the passage of only the amount of fluid necessary to maintain the preset pressure value. Since the pressure-reducing valve provides no flow path for the main flow to the reservoir, an external path must be provided to return pilot fluid to reservoir. The omission of such a drain path will permit a "pressure lock" that will render the pressure-reducing valve nonfunctional and leave it in a continuous open condition. This valve may be obtained either with internal check valve, for bidirectional flow, or without internal check valve, for unidirectional flow.

Fig. 5-5 depicts a reduced-pressure circuit to accomplish the same pressure versus force balance previously described in Fig. 5-2. The maximum pressure setting for the head-end chamber of cylinder C is established by the setting of relief valve D. The reduced pressure to the cap-end chamber is accomplished by pressure-reducing valve B in the conduit line to that chamber. The presence of check valve CV permits unaffected return flow from the cap-end chamber for return of the cylinder C piston. Unlike the Fig. 5-2 circuit, the full pressure value established by relief valve D is available to the other portions of the circuit indicated by connection Y, even during the forward or reduced-pressure stroke of cylinder C.

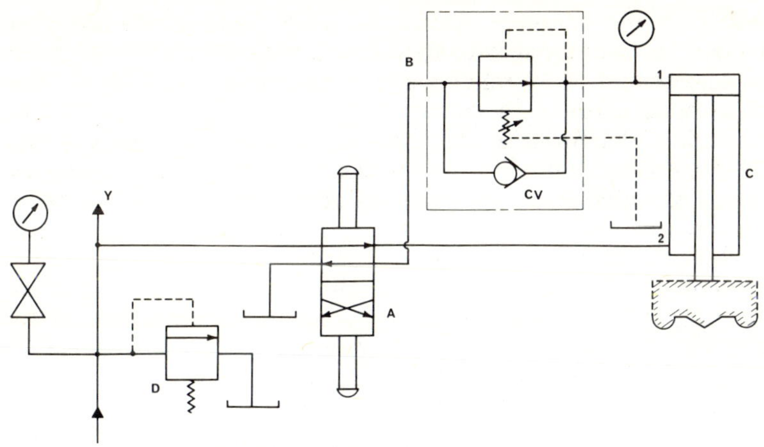

Fig. 5-5. A reduced-pressure circuit to accomplish pressure vs. force balance.

Fig. 5-6 illustrates a multiple-function circuit employing pressure reducing valve E to establish a reduced pressure in the cylinder D portion of the system. Cylinder C might be performing a clamping or holding function requiring the full pressure capability of the pump to securely clamp the workpiece. Cylinder D might be supplying drillpoint thrust and require

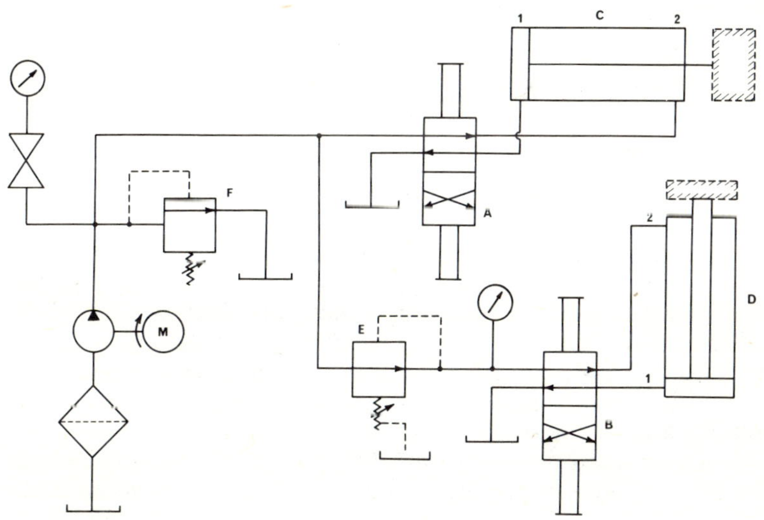

Fig. 5-6. A multiple-function hydraulic circuit using two pressure-reducing valves.

less operating pressure. As valve A is shifted, fluid is delivered to the cap-end chamber of cylinder C with maximum pressure value established by relief valve F. As valve B is shifted, fluid at a reduced pressure value established by pressure-reducing valve E is delivered to the cap-end chamber of work cylinder D. Reversal of valve B returns the cylinder D piston, and reversal of valve A returns the cylinder C piston. By virtue of the pressure-reducing valve being located in the supply line to directional control valve B, the same reduced pressure is presented to both cylinder D chambers alternately. Fluid exhausting from either cylinder D chamber is directed to the reservoir directly from the directional control valve via its tank port. No reverse flow is directed through the pressure-reducing valve; therefore, no check valve is required.

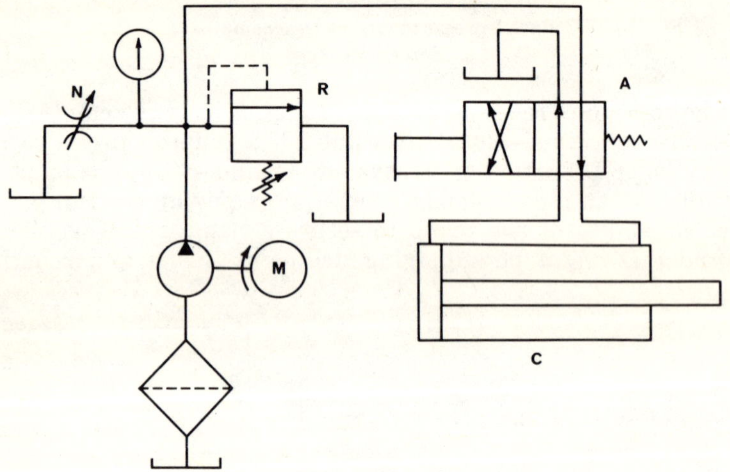

Fig. 5-7. Establishing maximum pressure with bleed off.

The reduced pressure beyond the pressure-reducing valve is not affected by pressure changes preceding it, as long as the inlet pressure exceeds the reduced pressure. This ability of the reducing valve to ignore fluctuating supply pressure is due to the fact that its pilot source is taken from downstream, as indicated by the graphic symbol. The drain line is indicated as terminating in the reservoir above fluid level.

A seldom-used method of establishing maximum pressure is the bleed-off system depicted in Fig. 5-7. Relief valve R establishes the maximum pressure value in the system, and noncompensated needle valve N can be regulated to establish any pressure value below that maximum. As valve A is shifted against its spring, fluid is directed to the cap-end chamber of cylinder C, causing its piston to extend and contact the work. With needle valve N adjusted to pass the full pump delivery to reservoir, no pressure will be realized in the system. As the needle valve orifice is reduced, full pump volume will still pass through it to reservoir, but the resistance to that volume of flow will increase. A gradual reduction of the needle valve orifice will create a gradual pressure rise until the maximum

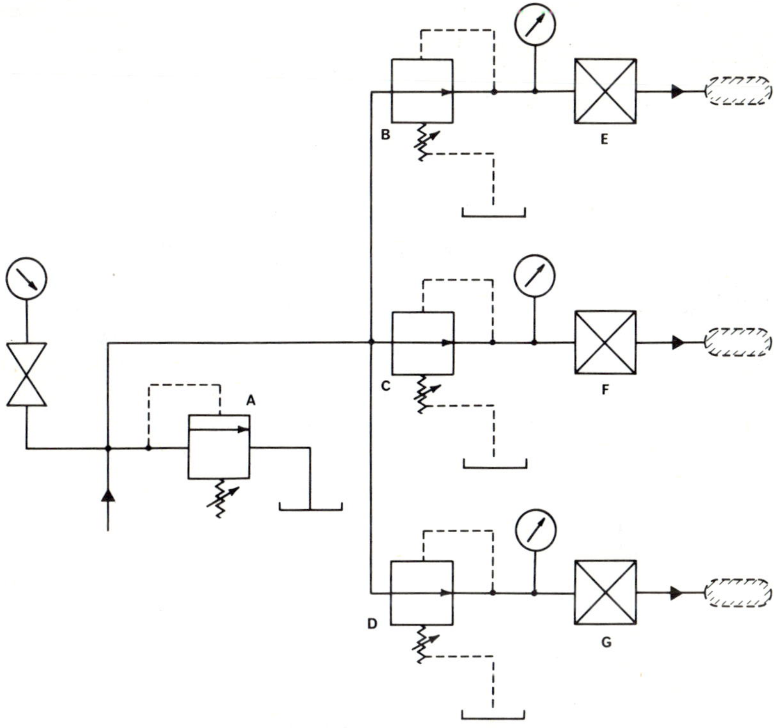

Fig. 5-8. Method of providing multiple pressures in a single system.

pressure value established by relief valve R is reached. Extremely fine resolution of pressure control can be obtained with excellent modulation characteristics. However, if a specific pressure value is to be maintained for any length of time, a constant fluid temperature must be maintained. Any rise in fluid temperature will lower the viscosity of the fluid and thereby reduce the flow resistance of a given orifice size.

In test applications, a variety of pressure selections is a rather common requirement. Fig. 5-8 depicts a simple method of providing multiple pressures in a single system. Relief valve A establishes the maximum pressure in the system. Pressure-reducing valves B, C, and D each establish a secondary reduced pressure value for their own stations. Shutoff valves E, F, and G are closed, except when an attached item is being tested. If a draining action following test is required, E, F, and G can be three-way valves with a return to reservoir provided for evacuating the test item. Valves E, F, and G could also be electrically operated to facilitate remote or automatic operation. Whichever form is desired should be so indicated on the circuit diagram by employing the appropriate symbol.

PRESSURE-COMPENSATED PUMP

An increasingly popular method of limiting the pressure in a hydraulic system is the use of a variable-volume, pressure-compensated pump. Pressure has been described as resistance to flow. The preceding examples in this chapter have used the approach of providing an alternate path to reservoir for the fluid being displaced by the pump. The variable-volume, pressure-compensated pump responds to increasing flow resistance by "destroking," or reducing its volumetric displacement. When a predetermined pressure value has been reached, the pump reduces its displacement to deliver only sufficient fluid

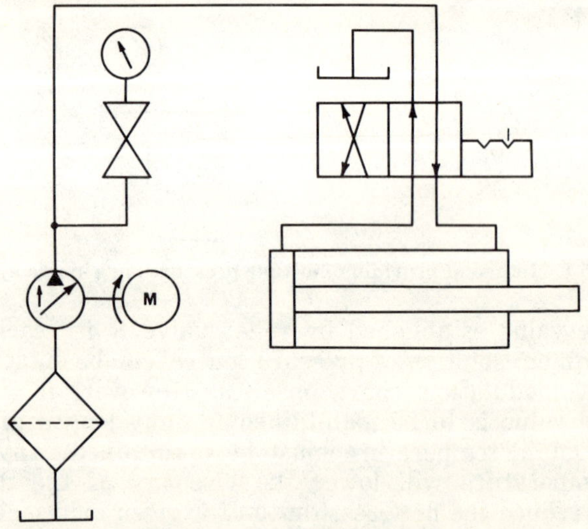

Fig. 5-9. Limiting pressure with a variable-volume, pressure-compensated pump.

to maintain that pressure. Since no alternate path to reservoir is required, the need for a relief valve has been eliminated. Fig. 5-9 depicts a simple circuit to illustrate the components required to complete a circuit utilizing such a pump for pressure-volume control.

INTENSIFIER CIRCUIT

Sooner or later the circuit designer will encounter a situation that calls for a relatively small volume of fluid at a pressure value appreciably higher than that required in the remainder

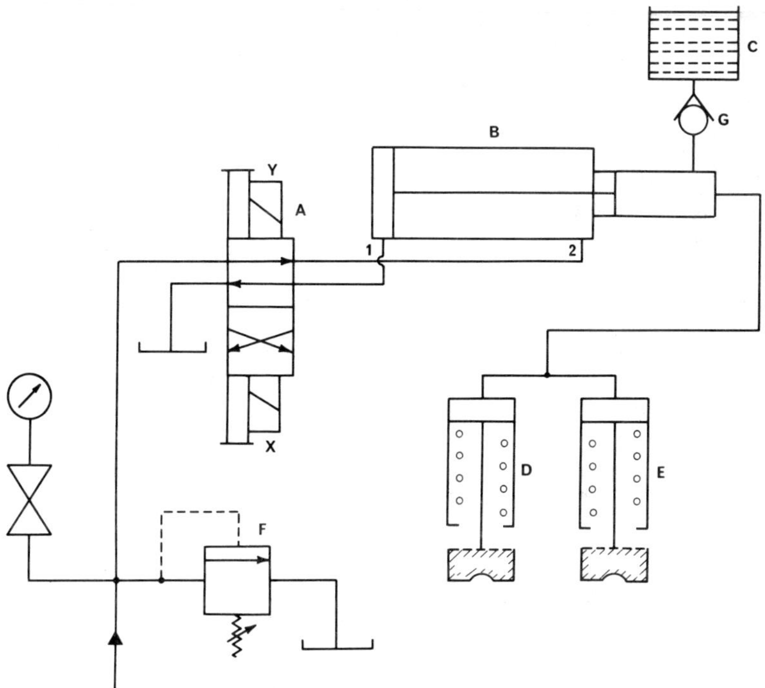

Fig. 5-10. A pressure intensifier circuit.

of the system. An existing system will normally have certain parameters already established, such as maximum pressure, maximum volume, and input horsepower. If the pump is a fixed-displacement unit or is limited in its pressure capacity, it may not be possible or practical for the designer to obtain the required pressure increase within the existing parameters and components. The intensifier or booster permits the designer to utilize existing volume by converting it to pressure, and thereby to attain a pressure value far in excess of the limits established by the pump-motor combination. Fig. 5-10 illustrates such a circuit.

The maximum pressure value in the existing system is established by relief valve F. Four-way directional control valve A controls the stroke direction of intensifier B. When solenoid X is energized, valve A shifts to deliver fluid to port 1 of intensifier B to activate its power stroke. As the power piston of the intensifier advances, fluid is displaced by the smaller piston in the high-pressure chamber. This displaced fluid is delivered to the cap-end chambers of single-acting, spring-return clamp cylinders D and E.

The maximum pressure value of this fluid is increased above the system pressure by the ratio of piston areas in the intensifier. For example, a 4-inch power piston with an effective area of 12.57 square inches might be working against a smaller piston of approximately one-inch square area. The ratio would then be approximately 12:1. This ratio means that the system pressure would be increased to twelve times its initial value in the intensified portion of the circuit.

When the work is completed, solenoid Y is energized and valve A returns to its initial position to power the return stroke of the intensifier B pistons. The springs in the clamp cylinders return the clamp cylinder pistons and force the fluid back into the high-pressure chamber of the intensifier. Any fluid that has been lost due to leakage will be replaced by the intensifier as it pulls fluid through check valve G from make-up chamber C, after clamp cylinders D and E have completed their return strokes. Check valve G prevents system fluid from escaping to reservoir C. The system is efficient and effective and can be a very useful tool for the imaginative circuit designer.

CHAPTER 6

Multiple-Action Circuits

Where multiple-cylinder or multiple-force components are employed in a circuit, a single four-way, directional control valve often is adequate. In order to obtain multiple functioning from such a valve, a simple method of redirecting the flow paths from it is necessary. The sequence valve offers an efficient means of accomplishing this redirection of flow.

In Fig. 6-1, it is desirable to have cylinder A complete its extension stroke before cylinder B begins its extension stroke. The use of sequence valve C in this illustration creates a delay of fluid delivery to the cap-end chamber of cylinder B, until cylinder A has completed its extension stroke and created a pressure buildup due to flow resistance. This pressure rise supplies a pilot signal of sufficient magnitude to pilot sequence valve C to its open position, thus redirecting the flow from valve D to the cap-end chamber of cylinder B to power its extension stroke. Reversal of valve D powers the retraction strokes of both cylinders simultaneously. The check valve in the sequence valve package permits free flow from the cap-end chamber of cylinder B, thus rendering sequence valve C nonfunctional during the retraction strokes of the two cylinders. The sequencing of the extension strokes of the two cylinders is accomplished by a pressure-sensing signal response.

Fig. 6-2 is similar to Fig. 6-1, except that an additional pressure-sensing sequence function has been added. In this illustration, the piston of cylinder A completes its extension stroke and reflects a pressure rise to pilot sequence valve C open. Fluid is thus redirected to the cap-end chamber of cylinder B to power its extension stroke. When four-way, directional control valve D is reversed, fluid is delivered to the head-end chamber of cylinder B to power its retraction stroke. When the cylinder B piston has fully retracted, a pressure signal of sufficient magnitude pilots sequence valve E to open and thus redirect the fluid flow to the head-end chamber of cylinder A. Its piston completes its retraction stroke, thus completing the cycle.

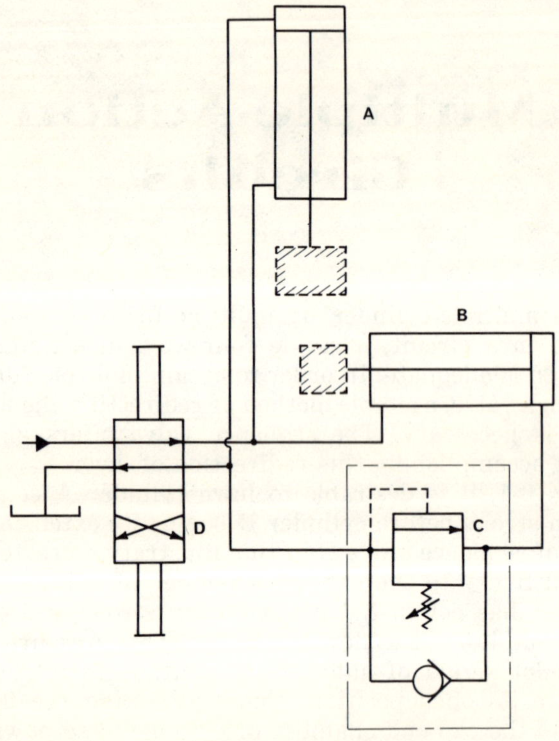

Fig. 6-1. Use of a sequence valve to delay cylinder extension.

Sequentially, the functioning is as follows:

1. Extension stroke of cylinder A piston
2. Extension stroke of cylinder B piston
3. Retraction stroke of cylinder B piston
4. Retraction stroke of cylinder A piston.

In a situation where A is the clamp cylinder and B is the working cylinder, the net result is a clamping of the part, with the extension and retraction of the work-cylinder piston occurring

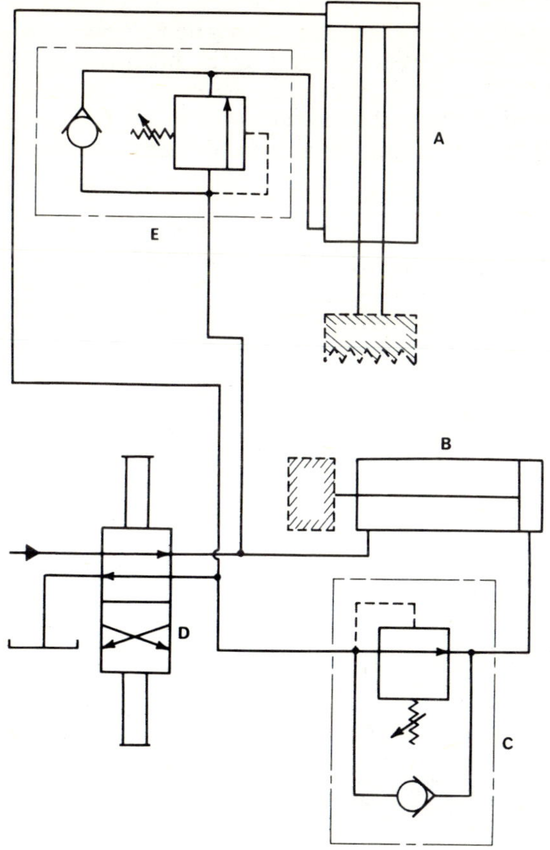

Fig. 6-2. Use of two sequencing valves to control cylinders.

prior to the unclamping of the part. The retraction stroke of cylinder B is triggered by manually revising four-way, directional control valve D, with the retraction stroke of cylinder A sequencing by the pressure-sensing signal to sequence valve E.

Fig. 6-3 offers a variation of the Fig. 6-1 circuit by employing pressure-sensitive switches to convert the pressure rise signals to electrical signals. The double-acting cylinder is controlled by a four-way, two-position, double-solenoid, directional control valve. Momentary actuation of push-button switch PB will energize solenoid A, shifting the directional control valve to a position that will deliver fluid to the cap-end cylinder chamber, powering the extension stroke of the cylinder piston. As the stroke is completed, pressure builds up to a value sufficient to trigger pressure switch PS. An electrical signal is then sent momentarily to solenoid B of the four-way, directional control valve, thus returning the control valve to its initial position for delivering fluid to the head-end cylinder chamber and powering the retraction stroke of the cylinder piston.

Fig. 6-4 duplicates the action of Fig. 6-2, while employing the same pressure signal to electrical signal conversion feature

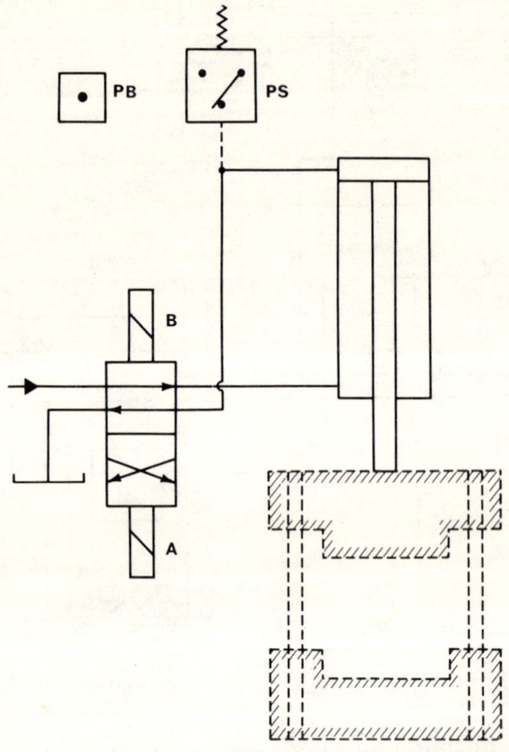

Fig. 6-3. Use of pressure switch to sequence cylinder piston.

of the pressure switch. Each cylinder is controlled by a double-solenoid, four-way, directional control valve. The sequencing signals are originated by four pressure switches. The desired sequence is:

1. Extension stroke of cylinder C piston
2. Extension stroke of cylinder D piston
3. Retraction stroke of cylinder D piston
4. Retraction stroke of cylinder C piston.

The cycle begins with the actuation of the start push button, which energizes solenoid 1 of valve A (line 1 in Fig. 6-4B).

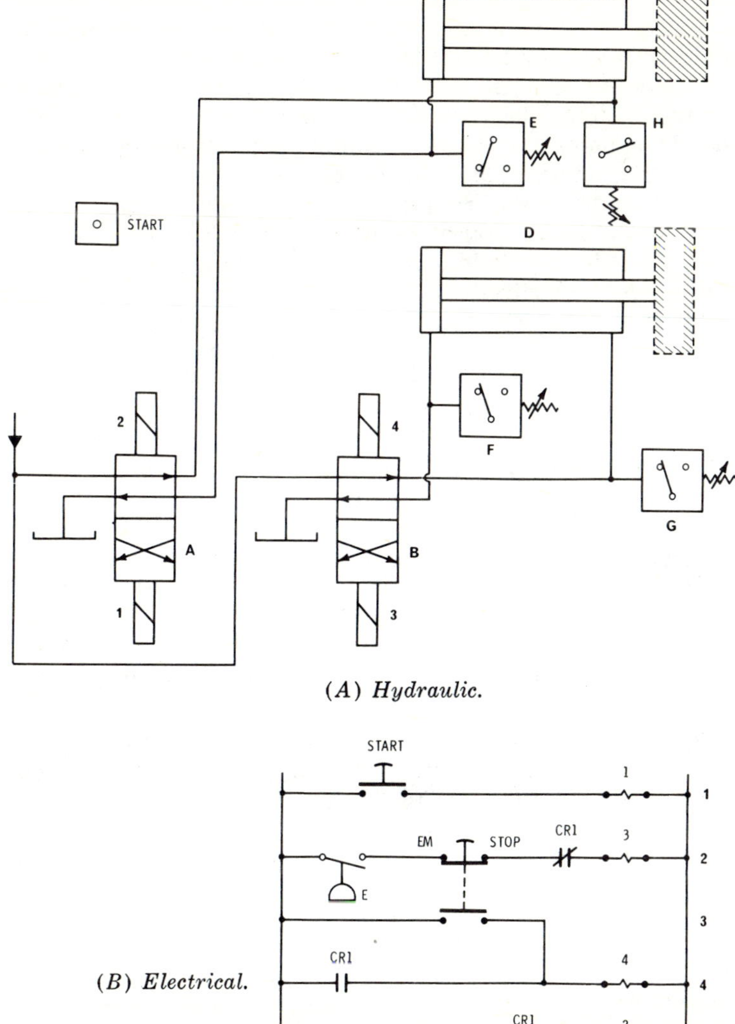

(*A*) *Hydraulic.*

(*B*) *Electrical.*

Fig. 6-4. Sequencing of two double-acting cylinders by use of four pressure switches.

This causes fluid to be delivered to the cap-end chamber of cylinder C, powering the extension stroke of its piston. When the pressure buildup triggers normally open pressure switch E, an electrical signal is sent through a normally closed set of relay CR1 contacts (line 2 of Fig. 6-4B) to energize solenoid 3 of valve B. This delivers fluid to the cap-end chamber of cylinder D, causing its piston to complete its extension stroke. At the completion of this extension stroke, the pressure buildup triggers normally open pressure switch F, thus delivering an electrical signal to the coil of relay CR1 (line 7 of Fig. 6-4B). This relay coil is then held energized through a set of its normally open contacts (line 6) in series with normally closed pressure switch H (line 7). An electrical signal is then delivered through another set of normally open CR1 contacts to solenoid 4 of valve B (line 4). This shifts valve B, causing fluid to be delivered to the head-end chamber of cylinder D and powering the retraction stroke of its piston. At the completion of this retraction stroke, a pressure buildup triggers normally open pressure switch G (line 5), thus delivering an electrical signal through a set of normally open CR1 contacts to energize solenoid 2 of valve A. This returns valve A to its original position, delivering fluid to the head-end chamber of cylinder C and powering the retraction stroke of its piston. At the completion of this retraction stroke, the pressure buildup triggers normally closed pressure switch H (line 7), thus interrupting the holding current to the coil of CR1. As the control relay deenergizes, the signals to solenoids 4 (line 4) and 2 (line 5) are both interrupted. At this stage, the circuit is again in its idle state and ready for the start of another cycle.

The emergency-stop push button (lines 2 and 3) can be used to reverse the stroke of the piston of cylinder D, with the retraction stroke of the cylinder C piston following in its normal sequence. Should it be desirable for the emergency-stop function to return both cylinder pistons simultaneously, a second set of normally open contacts could be incorporated into the emergency stop switch for the purpose of supplying an electrical signal to solenoid 2 (line 5) of four-way, directional control valve A.

While such use of pressure switches offers no positional interlock signals, it does offer a convenient method of electrical switching without making mechanical switch mountings directly on a machine. If it is imperative to actually sense positioning rather than using pressure sensing of the cylinder strokes, attention should be given to the limit-switch technique of signaling.

LIMIT-SWITCH, SOLENOID-VALVE SEQUENCING

The use of solenoid-operated, directional control valves for multiple-action sequencing requires some familiarity with electrical symbols and ladder diagramming. The end result may appear to be quite complicated, but when developed in a logical fashion, the process is not difficult. The fluid portion of such a circuit is quite simple, as indicated by Fig. 6-5A. In this application, the following sequence of steps is to occur:

1. Cylinder 1 piston extends
2. Cylinder 3 piston extends
3. Cylinder 2 piston extends
4. Cylinder 1 piston retracts
   Cylinder 4 piston extends
   Cylinder 5 piston extends
5. Cylinder 6 piston extends
6. Cylinder 2 piston retracts
   Cylinder 3 piston retracts
7. Cylinder 5 piston retracts
8. Cylinder 6 piston retracts
9. Cylinder 4 piston retracts

The circuit must be capable of a single cycle from a manual start. The circuit must be capable of automatic restart at the completion of the cycle. Limit-switch interlocks must provide for the following conditions:

1. Cylinder 1 must be fully extended before step 2 begins (LS1).
2. Cylinder 3 must be fully extended before step 3 begins (LS2).
3. Cylinder 2 must be fully extended before step 4 begins (LS3).
4. Cylinders 4 and 5 must be fully extended before step 5 begins (LS4 and LS5).
5. Cylinder 6 must be fully extended before step 6 begins (LS6).
6. Cylinder 3 must be fully retracted before step 7 begins (LS7).
7. Cylinder 5 must be fully retracted before step 8 begins (LS8).
8. Cylinder 6 must be fully retracted before step 9 begins (LS9).
9. Cylinder 4 must be fully retracted before cycle can repeat (LS10).

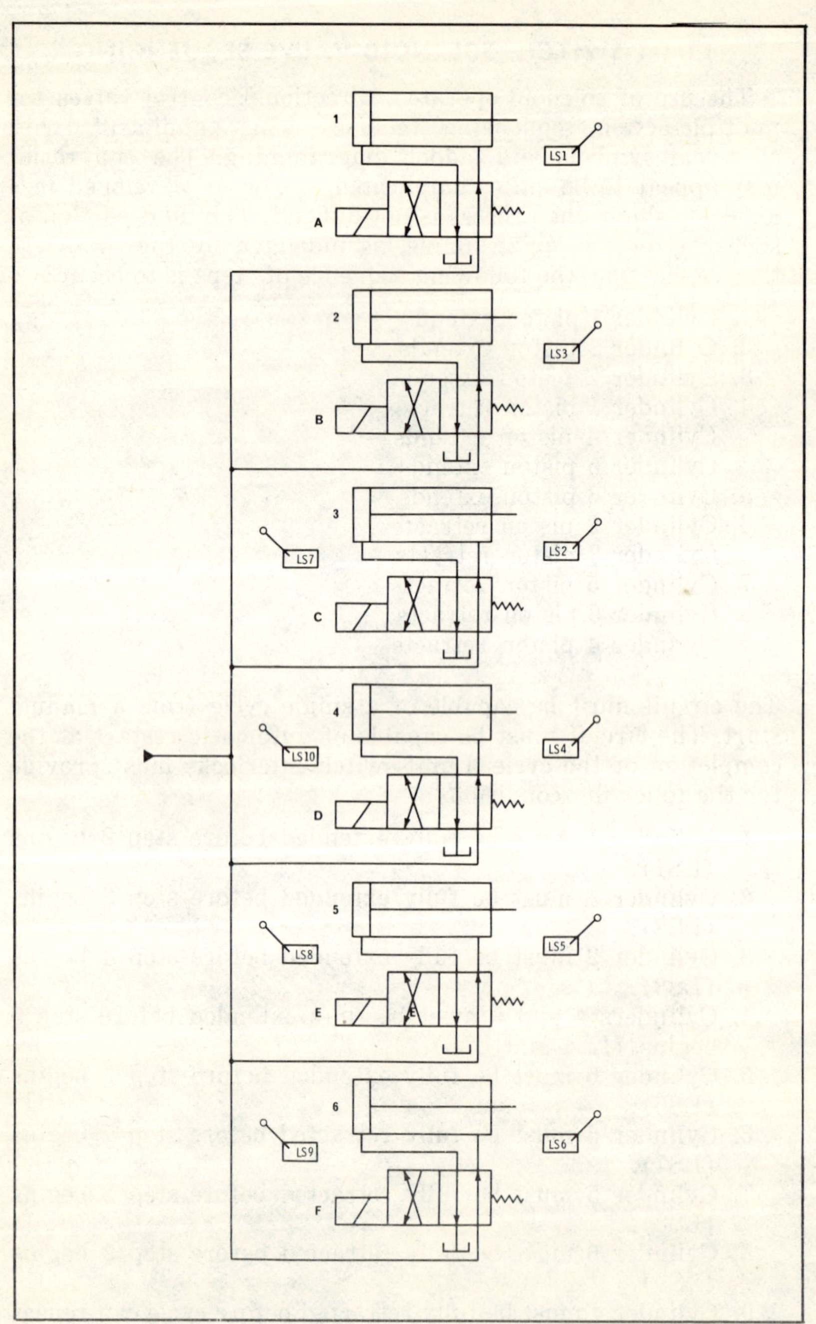

Fig. 6-5. Solenoid-operated control valves, for multiple sequencing.

To develop the electrical circuit from the foregoing requirements, the first step is to develop a truth table such as that shown in Fig. 6-6. The valve solenoids are listed A through F down the left side of the table. The sequence steps, 1 through 9, are listed across the top of the table. An *X* is placed in the square to the right of solenoid A in the step 1 column to indicate that this solenoid is to be energized at step 1. By referring to the sequence steps previously described, you will find that solenoid A must remain energized through step 3. Therefore, *X's* are placed in the squares for steps 2 and 3. Since valve A is not to be actuated throughout the rest of the cycle, *O's* can be placed in each of the other squares in the row for solenoid A. The squares for the other solenoids are marked similarly.

RELAY	VALVE SOLENOID	1	2	3	4	5	6	7	8	9
		PB OR LS10	LS1	LS2	LS3	LS4 & LS5	LS6	LS7	LS8	LS9
CR1	A	X	X	X	0	0	0	0	0	0
	B	0	0	X	X	X	0	0	0	0
CR2	C	0	X	X	X	X	0	0	0	0
CR3	D	0	0	0	X	X	X	X	X	0
CR4	E	0	0	0	X	X	X	0	0	0
CR5	F	0	0	0	0	X	X	X	0	0

Fig. 6-6. Truth table for development of electrical circuit of Fig. 6-7.

Directly beneath each step number is a space to indicate the switch that initiates that particular step. Since the cycle is to begin manually by operation of a push-button switch, "PB" is entered directly below step 1. The sequence requirements state that the cycle must be capable of automatic restart; therefore, LS10 (limit switch 10) is indicated as an alternate to the push button. A two-position selector switch labeled MAN-AUTO determines whether the sequence is a single-cycle, manual start, or automatic recycle through limit switch LS10. Each step is initiated by a switch or switches at the completion of the previous step.

Since the electrical signal from the push button is momentary, some means must be employed to sustain that signal through step 3 in order to keep cylinder 1 extended for the required portion of the cycle. This is most easily accomplished by the use of a relay (CR1). Limit switch LS1 presents the same problem; therefore, CR2 is employed to extend the signal of LS1. Since the signal requirement of LS2 is the same as its duration of actuation, no relay is needed to control valve B. Relays are required also for solenoids D, E, and F. The truth table now contains all the information necessary to develop an electrical ladder diagram. By converting this information to electrical symbols, the result is depicted in Fig. 6-7. The circled numbers indicate the step number to which each element relates. The MAN-AUTO selector switch is shown in the automatic recycle mode. For single-cycle, manual restart, the MAN-AUTO selector switch is actuated to its open position, this leaves limit switch LS10 performing only as an interlock function.

PROGRAMMING

One of the simplest means of accomplishing an automatic cycle is a technique known as *programming.* It is nothing more than the predetermined generation of command signals in a timed spacing of a sequential order. For the sake of simplicity, the same six cylinder and valve combination employed in Fig. 6-5 (ignore LS7, LS8, LS9, and LS10) can be sequenced by the programming technique. A total of six limit switches are used; one for each valve solenoid. These switches are actuated by a camshaft driven by a gear-reduction motor. The total time of a complete cycle is determined by the rate of rotation of the camshaft. The time between functions is determined by the relative rotational spacing of the cam lobes, and the duration of each signal is determined by the "length" of its cam lobe. Once set up, any sequence or timing can be altered by changing the rate of rotation of the camshaft, or by changing the relative positions or lengths of the cam lobes.

The electrical ladder diagram (Fig. 6-8) is quite straightforward, with each valve solenoid being assigned its own limit switch. The order of functioning of this system depends entirely on the order in which the limit switches are actuated and deactuated by their respective cam operators.

Fig. 6-9 is a variation of the electrical portion of the circuit shown in Fig. 6-8. In this electrical portion of the circuit, time-delay relays TR1 through TR6 are employed instead of limit switches actuated by a motor-driven camshaft. On-delay relays are cascaded, with each timing relay triggering a succeeding timing relay. In this case, the sequential order is established by the relationship of the timers, and the time spacing between functions is determined by the adjustable time-delay

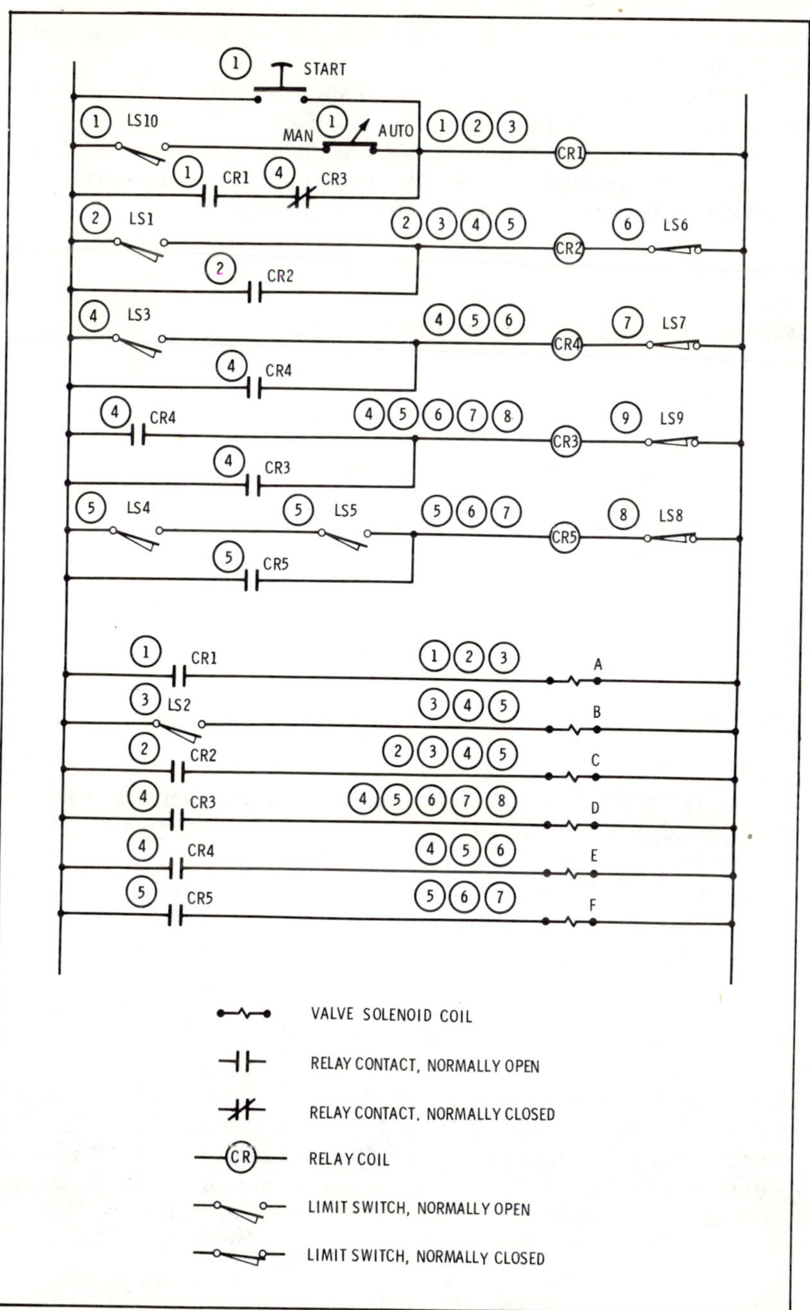

Fig. 6-7. Electrical circuit for controlling Fig. 6-5.

value of each timing relay. The programmed cycle is a one-cycle, manual-start sequence.

To start the cycle, the normally open push button in line 1 is actuated to supply power to the coil of control relay CR. The normally open CR contacts in line 2 immediately close and supply a holding signal to keep the control relay coil energized. The normally open CR contacts in line 3 immediately energize the coil of timing relay TR1. At the same time, the normally open CR contacts in line 9 close and energize the solenoid of valve A, causing the valve to shift and power the extension stroke of the piston of cylinder 1. When timing-relay TR1

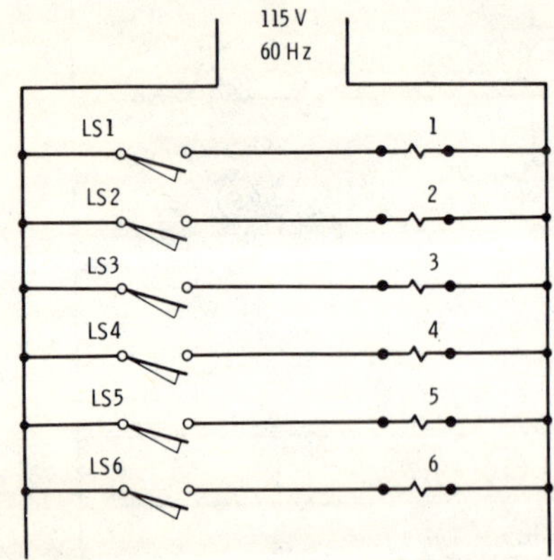

Fig. 6-8. Electrical circuit for programmed switching of Fig. 6-5.

completes its timeout, its normally open contacts in line 4 close, applying a signal to the coil of timing-relay TR2. The normally open contacts of relay TR1 in line 10 close and send a signal to the solenoid of valve B, causing valve B to shift and power the extension stroke of the cylinder 2 piston. As relay TR2 completes its timeout, its normally open contacts in line 5 close, energizing the coil of timing-relay TR3. At the same time, the normally open TR2 contacts in line 11 close, sending an electrical signal to the solenoid of valve C, thus powering the extension stroke of the cylinder 3 piston. When relay TR3 completes its timeout, the normally open TR3 contacts in line 6 close, energizing the coil of timing-relay TR4. At the same time, the normally open TR3 contacts in line 12 close, sending an electrical signal to the solenoid of valve D, thus powering the extension stroke of the cylinder 4 piston. When TR4 completes its timeout, the normally open TR4 contacts in line 7 close, energizing the coil of timing-relay TR5. At the same time, the normally open TR4 contacts in line 13 close, sending an electrical signal to the solenoid of valve E, thus powering the extension stroke of the cylinder 5 piston. When TR5 completes its timeout, the normally open TR5 contacts in line 8 close, energizing the coil of timing-relay TR6. At the same time, the normally open TR5 contacts in line 14 close, sending an electrical signal to the solenoid of valve F, thus powering the extension stroke of the cylinder 6 piston.

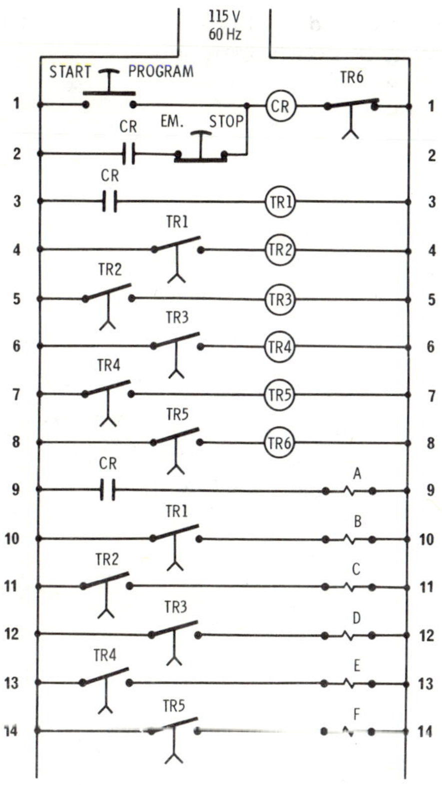

Fig. 6-9. Electrical diagram of time-delay relay programming.

As soon as timing-relay TR6 completes its timeout, its normally closed contacts in line 1 open, thus opening the circuit to the CR coil and resetting the cascade to an inactive condition. At this moment, the normally open contacts in lines 9, 10, 11, 12, 13, and 14 each open, breaking the signals to each of the six valve solenoids. As these valve solenoids are de-energized, each of the four-way, directional control valves returns to its spring-offset position, thus powering the return strokes of all six cylinder pistons simultaneously. If something other than simultaneous return of the six cylinder pistons is desired, the cascade could be continued by adding timing relays to interrupt solenoid signals and accomplish whatever sequence is desired. Automatic recycling could also be achieved by adding an additional timing-relay, or by using a limit switch at the retracted position of the last cylinder to return. A selector switch could then be employed in the same fashion as that in Fig. 6-7 to convert from signal cycle to automatic repeat. Should interlock protection be required between any two motions, limit switches can be added to any or all of the last six lines of the ladder diagram to accomplish this.

CHAPTER 7

Circuits for Increasing Speeds of Cylinder Travel

In any hydraulic circuit, the combination of volume and pressure determines the input horsepower requirement. Since maximum forces and maximum speeds are often not required simultaneously, many circuit variations can be employed to provide greater speeds during those times in the cycle when maximum forces are not needed, and lower speeds when maximum forces are needed. By effecting such variations in a circuit, far greater efficiencies can be obtained from the power supply factor available.

REGENERATIVE CIRCUITS

Fig. 7-1 illustrates the basic principle of the regenerative circuit. This circuit, in its idle state, is delivering fluid to the head-end chamber of cylinder B. When valve A is shifted to deliver fluid to the cap-end chamber of cylinder B, the fluid returning from the head-end chamber is channeled back into the supply feeding valve A, and thus back into the cap-end chamber. Assuming that the piston rod area is one-half the area of the cylinder piston, the forward speed of the cylinder piston will be twice that which could be achieved by the pump volume alone. As such, this is not entirely a working circuit, inasmuch as the net thrust of the cylinder is equal only to the piston rod area force. To realize the full force capability of the cylinder, attention should be given to the circuit in Fig. 7-2.

The circuit depicted in Fig. 7-2 is similar to that shown in Fig. 7-1, except that a second valve (B) has been added to the head-end cylinder line. Valve A and valve B are three-way, two-position, manually operated, detented valves. When valve A is actuated, the circuit performs in the same fashion as the circuit depicted in Fig. 7-1. The cylinder piston extends at

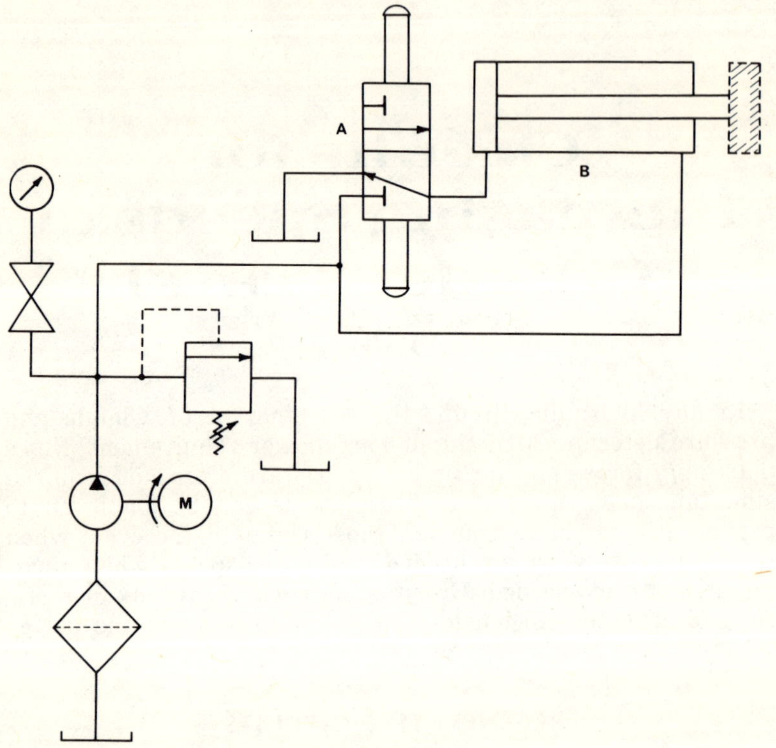

Fig. 7-1. Basic regenerative circuit.

twice its normal speed, but with only one-half its force effective. Upon contacting the work, valve B is actuated. This isolates the head-end chamber from the power portion of the circuit and connects it directly to the reservoir. As soon as this dumping is accomplished, the full piston area of cylinder C is utilized to deliver full cylinder force against the work. With both valve A and valve B returned to their initial positions, the cylinder piston retracts. The net result of this circuit is a doubling of the extension speed of the cylinder piston, with full power available when it is required to perform work at or near the end of the extension stroke.

The performance of the Fig. 7-2 circuit can be accomplished with a single valve as depicted in Figure. 7-3. The control valve in this circuit is a three-position valve with a center flow pattern combining both cylinder port A and cylinder port B with the power supply port. When the position 3 flow pattern is in effect, the cylinder C piston retracts. When the valve is shifted

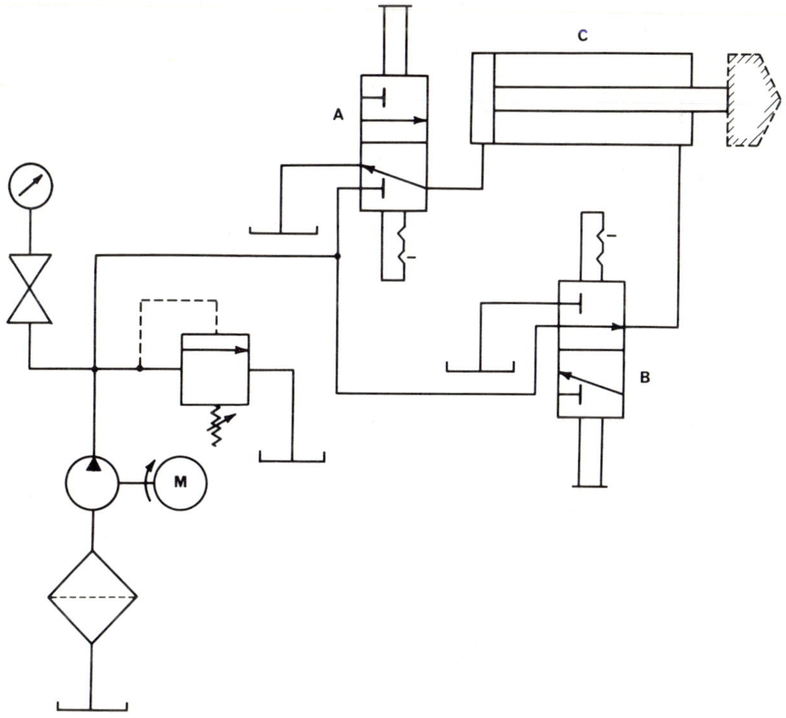

Fig. 7-2. Regenerative circuit using two valves.

to the center or number 2 position, the cylinder piston extends at twice normal speed but with half power, thus effecting the regenerative function. When the valve is shifted to the flow pattern of position number 1, the regenerative effect is nullified, and full cylinder power is utilized.

In each of the preceding regenerative circuits, the decision to nullify the regenerative pattern and effect full cylinder force is at the option of the machine operator. Fig. 7-4 depicts a circuit in which the decision to nullify the regenerative pattern is automatic. The forward stroke of the cylinder C piston is re-

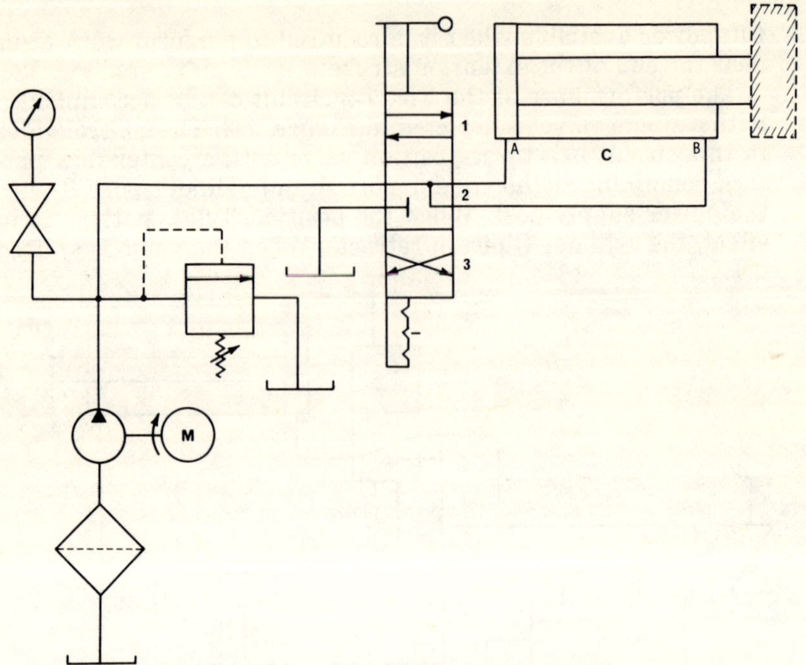

Fig. 7-3. Regenerative circuit using a three-position valve.

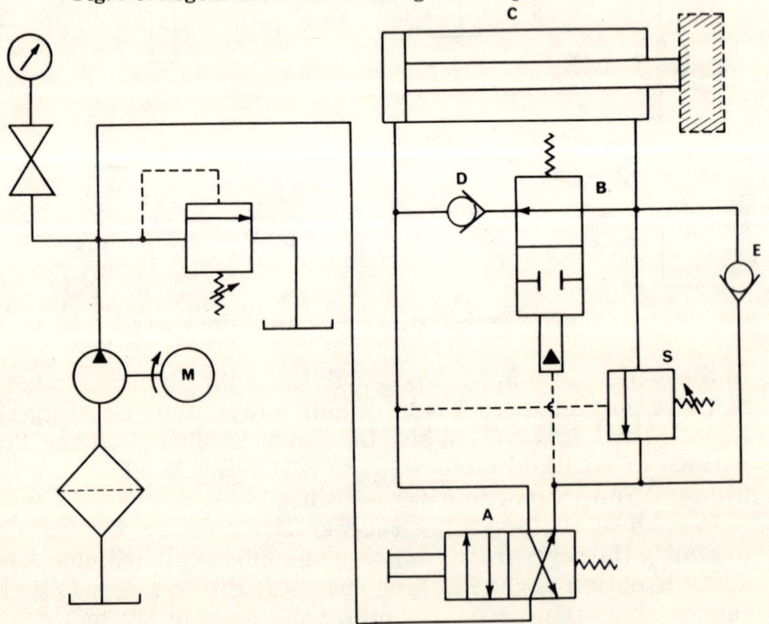

Fig. 7-4. Regenerative circuit with automatic nullifying.

generative until the work resistance is met. With valve A actuated, pressure supply is delivered to the cap-end chamber of cylinder C. Sequence valve S is closed, as is also check valve E. Fluid exhausting from the head-end chamber of cylinder C is delivered through two-way, normally open, pilot-operated valve B and through check valve D to join with the pump flow to the cap-end chamber. When the piston meets the work load, the pressure rise in the cap-end cylinder line pilots sequence valve S to its open position, thus causing the head-end cylinder chamber to be connected to the reservoir through control valve A. Check valve D also seats, thus delivering full power to the piston rod for accomplishing the work. As valve A is returned

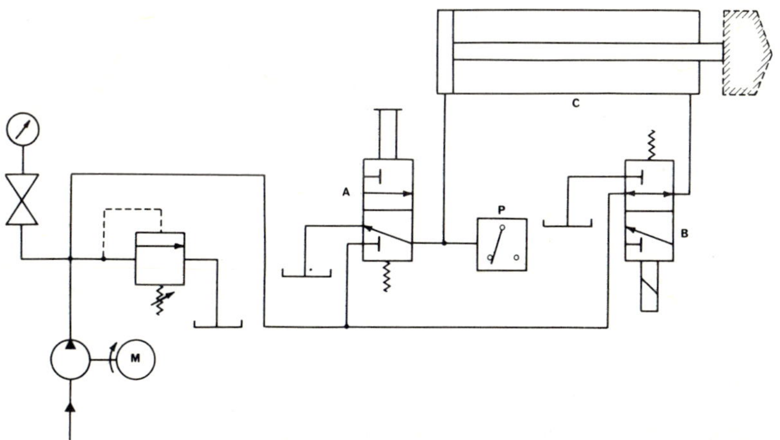

Fig. 7-5. Automatic nullifying with pressure switch.

to its normal spring-offset position, two-way, normally open valve B is shifted by pilot pressure to its closed position, and fluid is delivered directly through check valve E to the head-end chamber for the return of the cylinder piston.

Fig. 7-5 offers a second automatic regenerative system similar to that shown in Fig. 7-4. In this circuit, a pressure switch is used to actuate valve B to nullify the regenerative function. As valve A is shifted to the spring-offset position, fluid is delivered to the cap-end cylinder chamber, causing the piston rod of cylinder C to extend. Fluid leaving the head-end cylinder chamber passes through valve B to rejoin the pressure supply line ahead of valve A. As the cylinder C piston rod contacts the work, pressure builds up in its cap-end cylinder line, and pressure switch P closes, causing solenoid-operated valve B to energize. As valve B shifts to its energized position, the head-end cylinder line is connected directly to the tank, thus nullifying the regenerative effect. As valve A is returned to its spring-offset position, valve B also is de-energized, and fluid is directed to the head-end cylinder chamber to retract the cylinder C piston.

The preceding examples of regenerative circuits are by no means exhaustive; any combination of valving that serves the purpose of combining the two cylinder ports for the extension stroke of the cylinder piston can be productive.

PREFILL CIRCUIT

The closing speed of high-tonnage hydraulic presses with appreciable pretravel prior to contact with work can be greatly accelerated by the use of a prefill circuit. Fig. 7-6 illustrates such a circuit. Cylinder C is a large bore, ram-type, single-acting, hydraulic cylinder with smaller bore cylinder D attached to its piston through the cap-end cover. Cylinder D needs to be only of adequate bore to lift or return the cylinder C piston rod with its attached load. When the four-way control valve is shifted, fluid is delivered to the cap-end chamber of cylinder D, causing the cylinder D and cylinder C pistons to extend to meet the work. As the stroke of the cylinder D piston causes the cylinder C piston to advance, fluid is pulled from an overhead reservoir through pilot-operated check valve P to maintain a full cap-end cavity of cylinder C. At the forward extension of the cylinder C piston, with the cylinder C platen contacting the work, the pressure buildup in the cap-end line of cylinder D pilots sequence valve S open, thus permitting the full pump pressure to be delivered to the cap-end chamber of cylinder C. Check valve P closes, isolating the overhead reservoir from the circuit. At this point, the full effective area of the cylinder C piston is subjected to the maximum pressure, thus generating the full available force of the cylinder C piston. When the four-way, directional control valve is reversed, fluid is directed to the head-end port of cylinder D; sequence valve S closes; and pilot-operated check valve P is opened. Thus, as the cylinder D piston retracts, pulling with it the piston of cylinder C, the fluid in the cylinder C cap-end chamber is returned to the overhead reservoir. The head-end port of cylinder C is merely opened to the reservoir to permit the return to the reservoir of any fluid that might bypass the cylinder C piston seals. The cycle is thus complete. The net result is a rapid approach at a speed far in excess of the volumetric capacity of the hydraulic pump, with full force becoming available as the work load is

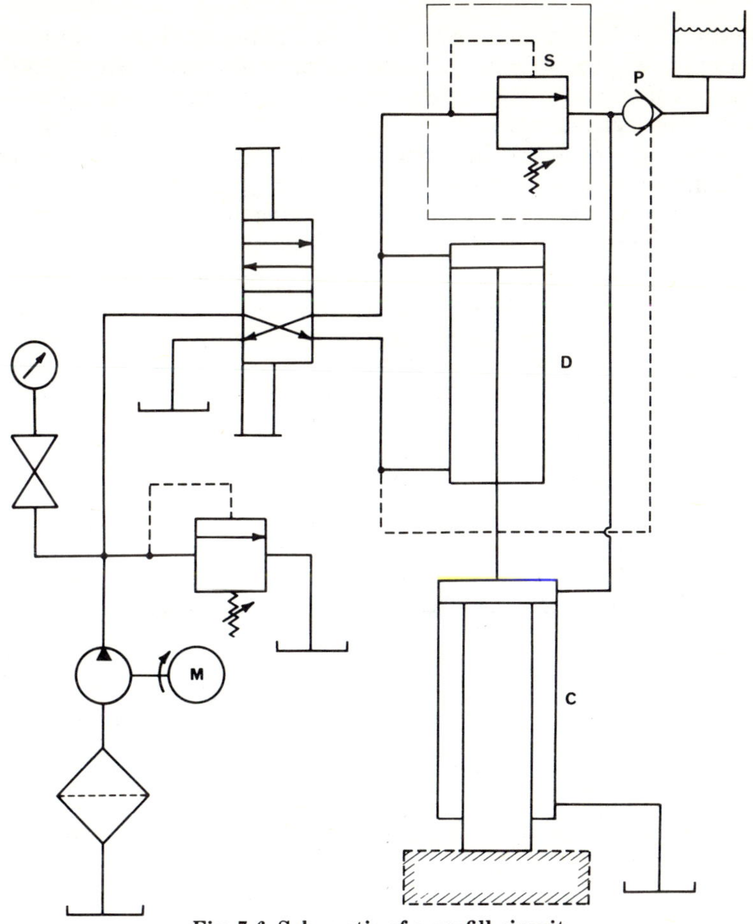

Fig. 7-6. Schematic of a prefill circuit.

contacted. The shift to full force is accomplished automatically when the pressure buildup triggers sequence valve S.

HI-LO CIRCUIT

When a hydraulic cylinder is required to move an appreciable distance at a comparatively high rate of speed, and with a high force available only at or near the end of the stroke, the Hi-Lo circuit lends itself readily to this requirement. Fig. 7-7 depicts a high-volume, low-pressure pump in combination with a low-volume, high-pressure pump to meet this requirement. In this circuit, the B pump is considered to be the high-volume, low-pressure pump, and A is considered the low-volume, high-pressure pump. When control valve E is shifted to deliver fluid to the cap-end chamber of cylinder C for high-speed travel, both pump A and pump B are delivering their volume to the circuit. When the cylinder C piston meets the resistance of the work, the pressure reflected in the supply line pilots unloading valve U to its open position, thus dumping the full volume of pump B directly to the reservoir. Check valve Y seats, completely isolating pump B from the high-pressure portion of the circuit. Pump A continues to deliver its fluid at high pressure to the system to accomplish the high-tonnage requirement upon contact with the work load. Relief valve R establishes the maximum pressure valve of the circuit. When the four-way control valve is returned to its initial position, fluid is delivered to the head-end chamber of cylinder C to cause the retraction of its piston. During the retraction stroke, both pump A and pump B are supplying fluid to the system. Upon the full retraction of

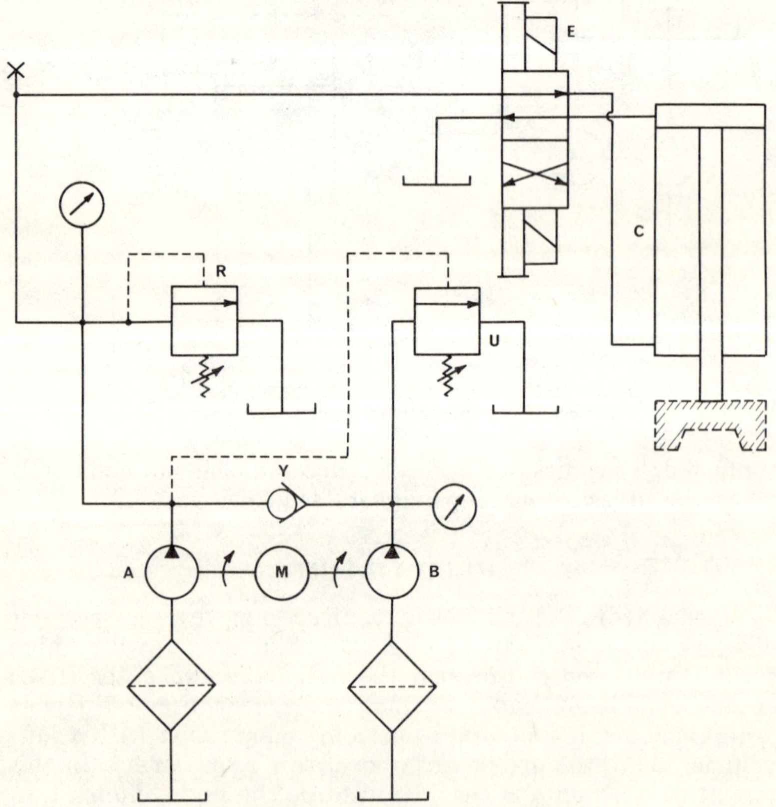

Fig. 7-7. Schematic of a Hi-Lo circuit.

the cylinder C piston, the unloading valve U is again piloted open, permitting high-volume pump B to unload directly to reservoir. High-pressure, low-volume pump A continues to deliver high-pressure fluid to the system, and its delivery is used to maintain the cylinder C piston in retracted position. Such a circuit permits a high volume of fluid to be delivered during the low-force portion of the circuit, but with a considerable savings in the amount of horsepower consumed.

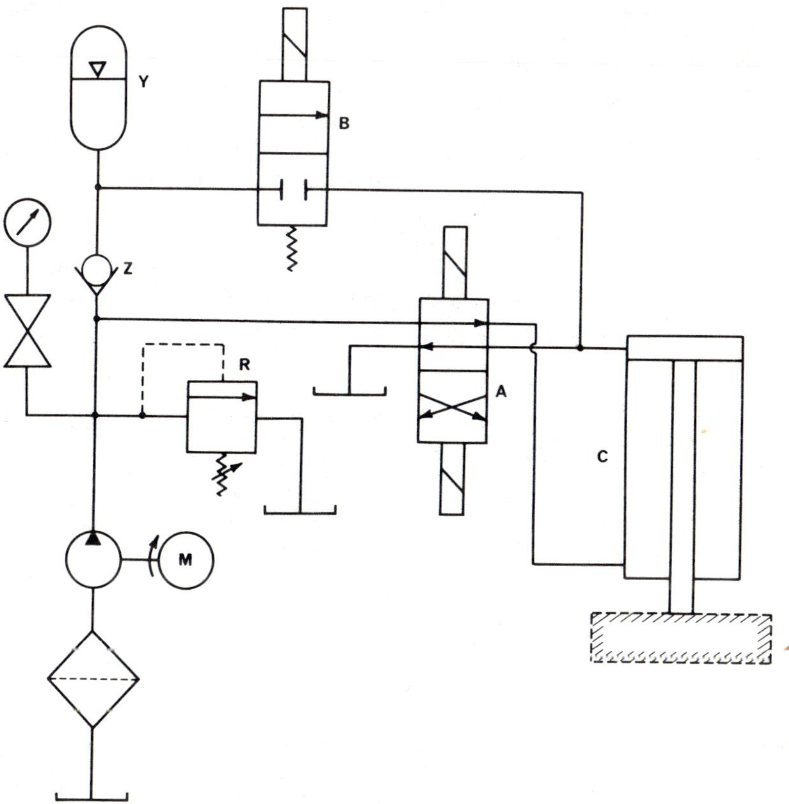

Fig. 7-8. Diagram to illustrate use of an accumulator.

ACCUMULATOR CIRCUIT

Whenever a hydraulic circuit is idle, such as for loading and unloading fixtures, an accumulator may be used as an excellent means of storing the energy being delivered by the hydraulic pump. By then calling for this stored energy in that portion of the cycle when greater speed is required, a greater percentage of the pump capacity is utilized.

Fig. 7-8 depicts an accumulator circuit that provides such a storage facility. During the idle time of the cycle, (when all valves are in the positions shown in the circuit) fluid is delivered through check valve Z to accumulator Y, until the pressure value of relief valve R is reached. When this pressure value is reached, the volumetric output of the hydraulic pump is spilled over the relief valve and back to the reservoir. By shifting both valve A and valve B, the output of the hydraulic pump and the fluid stored in the accumulator Y are delivered through two-way normally closed valve B to the cap-end chamber of cylinder C. As soon as the work is contacted, valve B can be de-energized, thus isolating accumulator Y from the circuit, permitting the full pressure capacity of the pump to be delivered to the cap-end cylinder chamber. As four-way control valve A is returned to its initial position, fluid is delivered to the head-end chamber of cylinder C to retract the cylinder C piston, thus completing the cycle. The relative functioning of valve A and valve B can be achieved by automatic electric control, or by manual selection, depending upon the specific requirements of the circuit.

CHAPTER 8

Circuits for Combating Heat

In a hydraulic system, the problem of heat generation is one that cannot be ignored. Heat not only causes hydraulic fluids to deteriorate, but it also changes the characteristics of the fluid and the performance of components within the system. Uncontrolled temperatures within a hydraulic system can either destroy or render ineffective many of the standard materials that are used in system components. Clearances of moving or mating parts made of dissimilar materials may be altered due to different coefficients of expansion. These changes may create excessive clearance flows or increased frictional resistance, resulting in the creation of more heat.

Most of the heat that is encountered in a hydraulic system originates at the pump unit. Heat is created or induced into the hydraulic fluid as a direct result of the work that is performed on the fluid in moving it against a pressure resistance. Additional heat is added to the fluid at points in the system where frictional resistance is encountered. Most hydraulic systems operate with greatest efficiency in a temperature range of 120°F to 140°F. When the fluid temperature begins to exceed approximately 170°F, maintenance problems should be anticipated.

The most common remedy for excessive temperatures in a hydraulic system is the heat exchanger. The two main types of heat exchanger used in industry are the air-transfer type and the water-transfer type. Some work requirements lend themselves to other means of controlling, or at least reducing, the heat buildup in a hydraulic system. By giving careful attention to circuit requirements, and taking advantage of a few heat-prevention principles, the heat buildup problem can often be eliminated, or at least alleviated.

PRESSURE-COMPENSATED PUMP CIRCUITS

The variable-displacement pump is one practical approach to the heat problem. Fig. 8-1 depicts a hydraulic circuit utiliz-

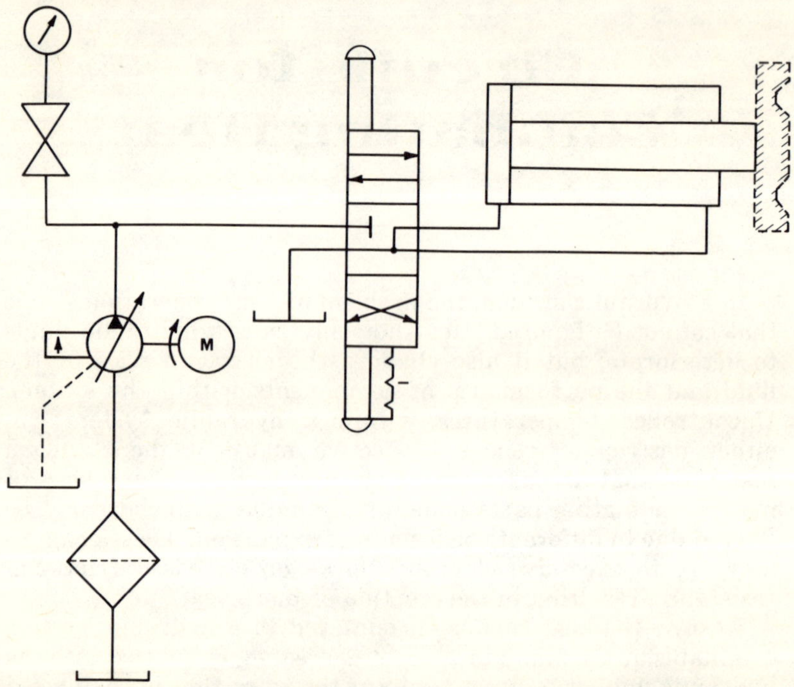

Fig. 8-1. Hydraulic circuit utilizing a variable-displacement pump.

ing such a pump. Since this type of pump reduces its volumetric pumping rate as the demand for fluid is satisfied in a system, less work is performed within the pump, and thereby less heat is added to the system. In the case of a molding press with a long curing cycle, for example, a variable displacement, pressure-compensated pump automatically adjusts itself to deliver only sufficient hydraulic fluid for maintaining the required pressure in the system.

UNLOADING CIRCUIT WITH DUMP VALVE

As previously stated, heat generation in a hydraulic circuit is due to work done on the fluid as a result of delivering a volume of fluid against a pressure resistance. Fig. 8-1 shows a reduction in the volume of fluid being delivered by reducing the volumetric displacement of the pump. In Fig. 8-2, the work performed on the fluid is reduced by reducing the pressure resistance to the volumetric displacement of a fixed-volume pump. The double-acting cylinder is controlled by four-way, two-position, double-solenoid valve V1. In the idle state shown in the circuit, no pressure is required to hold the cylinder in its retracted position. It is also true that in this retracted position, no work is being performed. Therefore, the pressure value established by the relief valve is not required at this point. Since the fixed-displacement pump is constantly delivering a volume of fluid, the problem of heat generation due to work performed on the fluid is eliminated by shifting valve V2 to its open position. This allows the oil to recirculate through

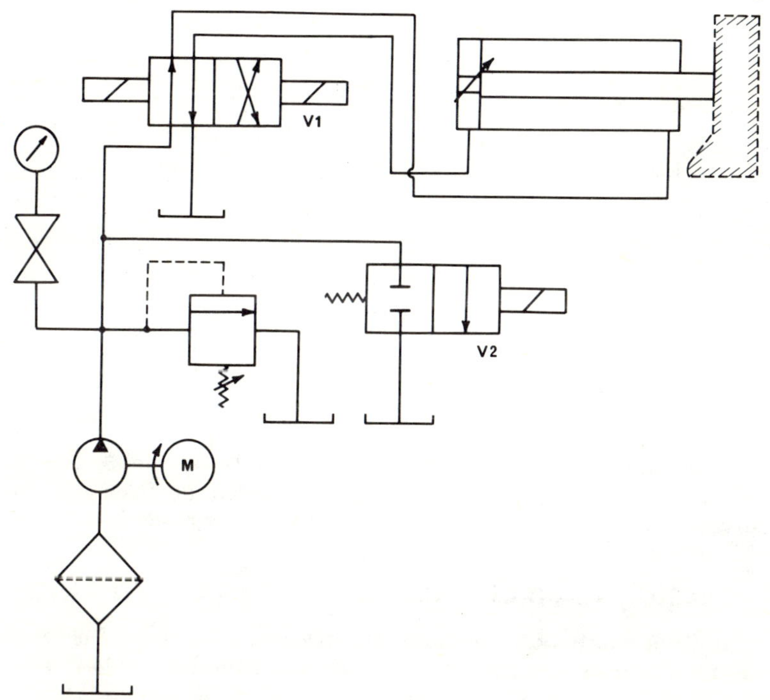

Fig. 8-2. Circuit using a normally open valve to control reduced pressure resistance with a fixed-volume pump.

open valve V2 and directly back to the reservoir with minimal pressure resistance. If pressure is normally required in the circuit, a two-way, normally closed valve, as indicated in Fig. 8-3, would be appropriate. In an application where no pressure is required for the greater percentage of the cycle time, two-way, normally open valve V2 could be employed. With this latter usage, the solenoid of valve V2 would need to be energized

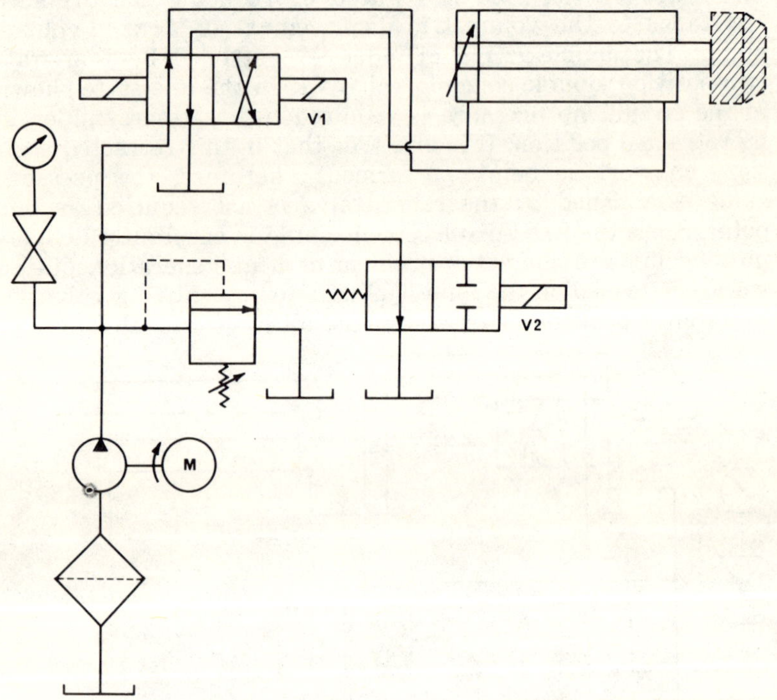

Fig. 8-3. Circuit using a normally closed valve to control reduced pressure resistance with a fixed-volume pump.

in order to achieve pressure within the circuit. Thus, the selection of normally open or normally closed is based on the percentage of time the solenoid would be energized.

UNLOADING CIRCUIT USING RELIEF VALVE VENTING

Another method of reducing the pressure loading of a fixed-volume pump is depicted in Fig. 8-4. A secondary valve is used in a fashion similar to that in Fig. 8-2; however, in this illustration the two-way valve is used to vent the pilot portion of the relief valve. This arrangement offers a possible component cost savings over the circuits in Figs. 8-2 and 8-3. The former circuits require a V2 valve of sufficient size to pass the entire volumetric capacity of the pump. The unloading feature of Fig. 8-4 requires a V2 valve with only sufficient capacity to pass the small volume of fluid carried by the pilot cap of the relief valve, usually not over 1 gpm.

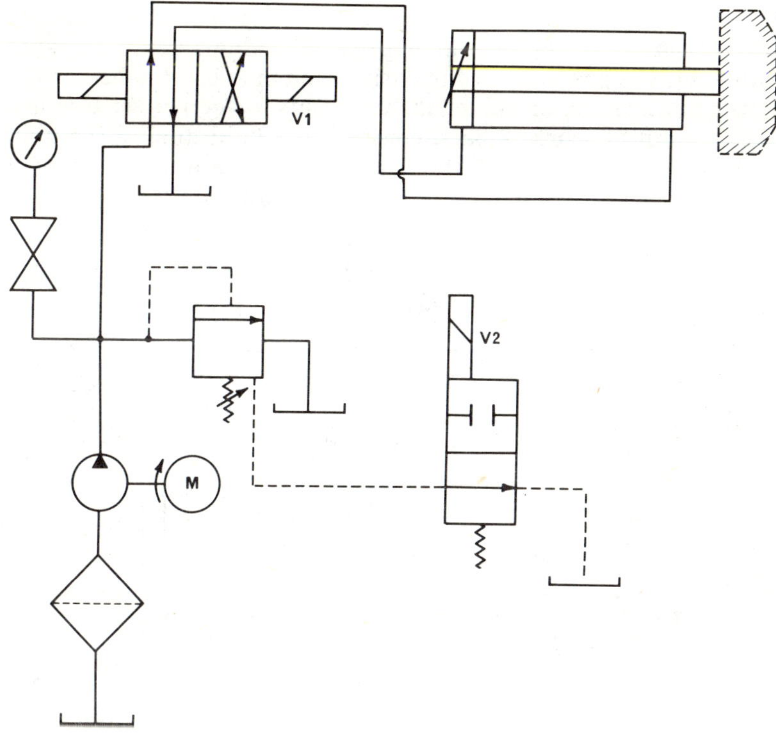

Fig. 8-4. Circuit using a vented relief valve to reduce pressure loading.

SECONDARY RELIEF VALVE CIRCUIT

Many single-cylinder valve hydraulic applications provide for work being performed only on the extension stroke of the cylinder. In such an application, the pressure requirement for the retraction stroke is only sufficient to return the cylinder piston from the work. Fig. 8-5 depicts a circuit that incorporates a secondary relief valve for accomplishing this lower pressure value on the nonworking stroke of the cylinder. A fixed-volume, constant-displacement pump supplies fluid to the system. The directional control valve is a four-way, two-position, hand-operated valve labeled V1. Relief valve R1 establishes the maximum pressure value in the system. This value is in effect for the full forward or work stroke of the cylinder. When four-way control valve V1 is shifted to the return position, fluid is delivered to the head-end port of the cylinder, causing the cylinder piston to retract. Pressure in this portion of the circuit is established by the pressure value setting of secondary relief valve R2. Since the lower value setting of relief valve R2 takes priority over the higher-pressure value setting of relief valve R1, the pressure in the entire system is established at the lower value of relief valve R2. With this lower pressure value, a minimum of work is performed on the fluid during the nonworking or idle time of the cycle. Thus, the amount of heat generation within the fluid is held to a minimum level.

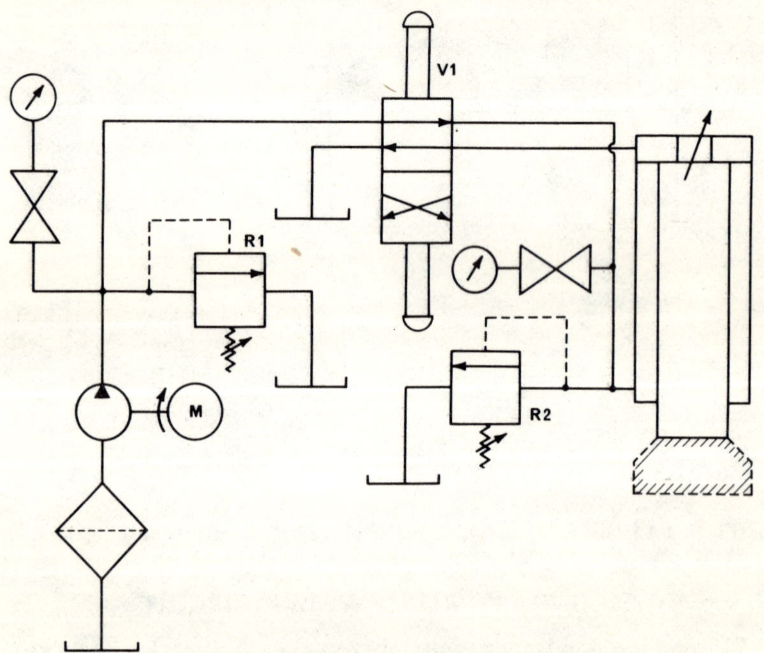

Fig. 8-5. Reducing pressure on nonworking stroke with secondary relief valve.

UNLOADING CIRCUIT USING TANDEM-CENTER VALVE

Removing the pressure resistance to the fluid volume of the pump was accomplished in Figs. 8-2 and 8-3 by the use of a

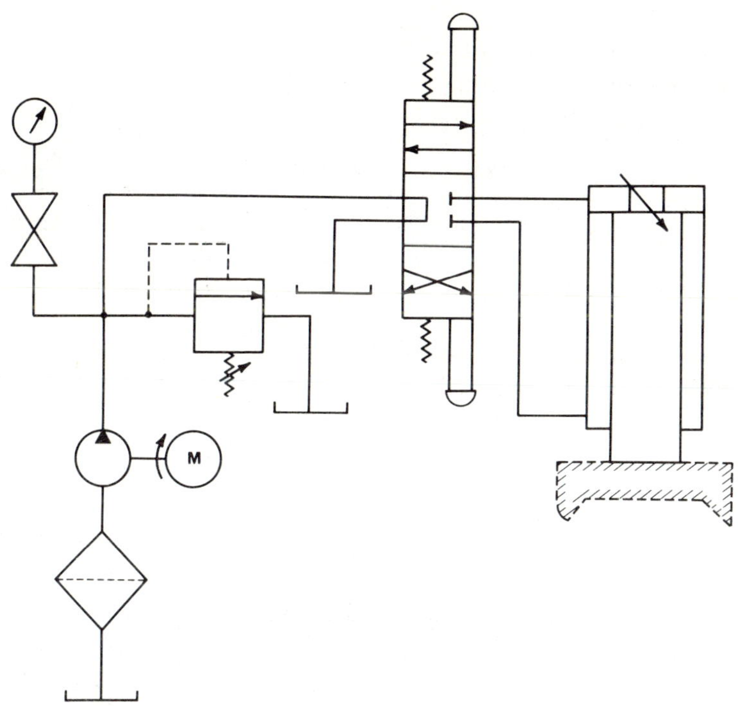

Fig. 8-6. Reducing pressure with a "tandem-center" valve.

secondary two-way valve providing a direct return flow path to the reservoir. This direct return flow path can also be supplied by a special configuration of the center position of a three-position, directional control valve. Fig. 8-6 depicts such a three-position valve with a center porting condition usually referred to as a "tandem center." The directional control valve used in this circuit is a manually operated valve. The operating pressure within this circuit is established by the pressure value setting of the relief valve. This pressure value is in effect for both the extension and retraction strokes of the cylinder. At any time during the cycle when no force or movement is required from the operating cylinder, the valve is returned to the center position. In this center position, the internal port paths provide a direct path from pump to reservoir. With the control valve in this center position, a very nominal pressure resistance is offered to the full fluid flow from the pump.

Fig. 8-7, a variation of Fig. 8-6, utilizes a three-position, tandem-center, double-solenoid, pilot-operated control valve. When such a control valve is employed, a check valve is placed in its tank line, providing sufficient back pressure in the system to supply pilot pressure for the functioning of the control valve. A check valve with a 65 psi cracking pressure is normally used for this purpose. A relief valve set for 65 psi could be employed instead of the check valve.

In Fig. 8-8, a different spool configuration is incorporated in the control valve, with a nominal pressure being applied to the head-end chamber of the vertical-mounted cylinder. The pressure value of the fluid in the head-end chamber is established by the cracking pressure of the check valve in the return to

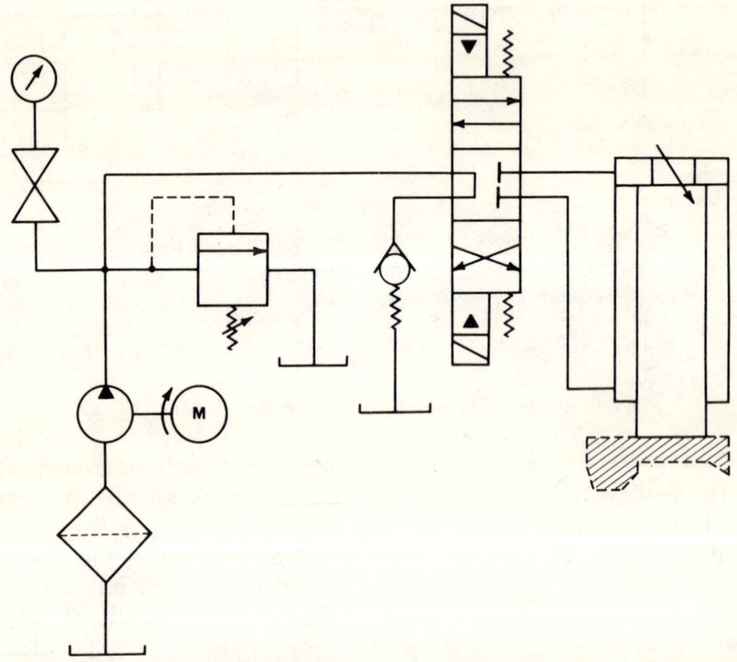

Fig. 8-7. Using a three-position tandem-center, double-solenoid, pilot-operated valve to reduce pressure.

tank line of the directional control valve. This supplies sufficient pressure to the head-end chamber of the cylinder to hold the cylinder piston in the retracted position and also supplies pilot pressure to the pilot-operated, directional control valve.

HI-LO SYSTEM USING AIR-OPERATED HOLDING PUMP

It is not unusual in a hydraulic press circuit used for rubber or plastic molding to encounter an extended cure time that requires high fluid pressure, but no appreciable volume of fluid. Since only the opening and closing of the press requires any fluid volume, the high-pressure portion of the requirement can easily be satisfied by incorporating a pressure demand, air-operated unit. Such a system is shown in Fig. 8-9. The travel portion of this circuit is satisfied by the use of the fixed-volume pump. The pressure value of this portion of the circuit is determined by the pressure value setting of relief valve R. The opening and closing of the press powered by cylinder A is controlled by four-way, two-position, double-solenoid, directional

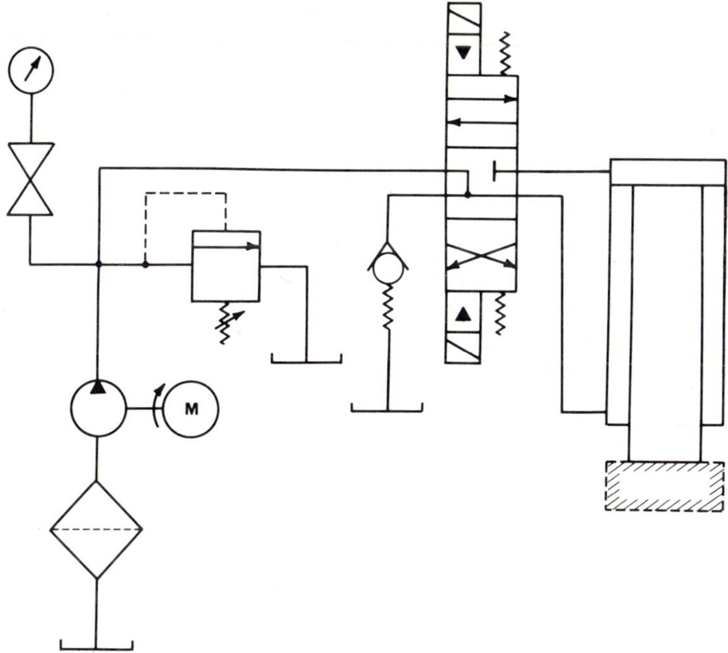

Fig. 8-8. Circuit similar to Fig. 8-7 using different spool configuration.

control valve V1. Closing of the press is initiated by momentary operation of solenoid S1 of valve V1. This causes the V1 valve to shift, delivering fluid at nominal pressure to the cap-end chamber of cylinder A. When cylinder A is in its extended position, a signal is then given to solenoid S-3 of air valve V2, thus supplying air to air-oil booster B. This boosts the pressure of the fluid in the cap-end chamber of cylinder A and holds this pressure throughout the cure cycle of the press. Check valve C in the pressure supply line of directional control valve V1 is closed, thus preventing the loss of the higher-pressure fluid through relief valve R. During this curing cycle, the fixed-volume pump is merely dumping its volume through relief valve R and back to the reservoir at a very low pressure. Should the system develop internal leakage that would make the volume of fluid available from the high-pressure chamber of booster B inadequate, the booster could utilize an automatic reciprocating valve arrangement that would recycle itself upon receiving any additional volume demand. The operating advantages of this system include the absence of any heat generation within the high-pressure holding portion of the circuit and the ability to utilize the horsepower stored in the compressed air system without actually consuming it.

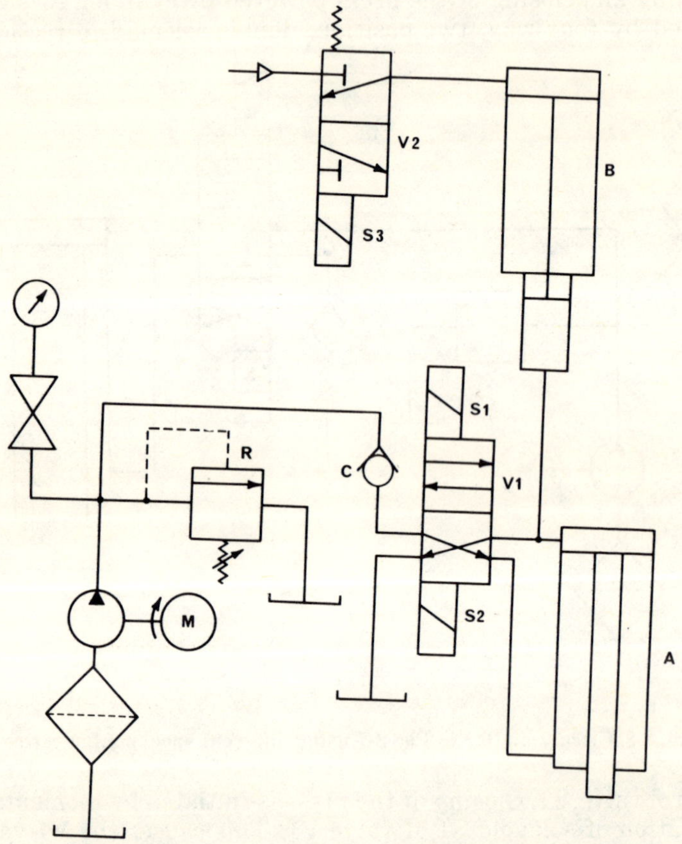

Fig. 8-9. Hi-Lo system using an air-operated holding pump.

CHAPTER 9

Circuits for Controlling Shock

In a hydraulic system, pressure is most accurately described as resistance to flow. The two hydraulic components employed for limiting maximum pressure are the relief valve and the pressure-compensated pump. These two components operate on different parameters of a circuit to control the flow resistance. In the case of the relief valve, the pressure limit, when reached, pilots the relief valve open to channel the fluid directly back to the reservoir. This maintains approximately the same constant volume that the pump is capable of delivering. The pressure compensator of a hydraulic pump reduces the volumetric displacement of the pump when a predetermined flow-resistance pressure level is reached. In summation, the relief valve offers an alternate flow path for the entire pump volume, while the pressure compensator reduces the volume of fluid being delivered by the pump.

Each of these two components has a time lag in its reaction to a pressure rise. This time lag is commonly referred to as "response time." Thus, a sudden pressure rise that occurs in less than the response time of either component can result in peak pressure rises of extremely short duration. Such pressure peaks are commonly referred to as hydraulic shock.

These reactive shocks are generated within a hydraulic system by sudden changes in the flow path, momentarily reducing the ability of the flow path to accept the full pump delivery. A cylinder suddenly reaching the full extension or retraction of its piston will bring this flow capacity of the conduit lines to a sudden stop. Also the shifting of solenoid-operated or pilot-operated control valves can create momentary stoppages within the normal flow paths. Sudden reductions or stoppages will invariably reflect a sudden rise in pressure due to the infinite and complete reduction of the flow path capacity at valve spool crossover points. The higher the normal operating pressure in a hydraulic system, the greater will be the intensity of these transient shocks. Such shocks are difficult to measure with standard gaging, due to the extremely short duration of their intensity. A gage might indicate the presence of some shock, but usually the magnitude of such shocks are far in excess of the ability of the gage to respond during the brief duration of the shock. Sophisticated measuring equipment has been used on many occasions to determine the magnitude of such transient shocks; readings in excess of 10,000 psi have been registered in 2000-psi systems. Discernable evidence of such shocks usually takes the form of jumping flexible conduit lines, an audible pounding effect, and a severe movement of components above and beyond those designed into the system. Such visible evidences of shock are not only unpleasant to observe, but are unquestionably damaging to the system. Such shocks can readily surpass design limits of components, resulting in inevitable, premature failure of such components. The alert circuit designer will anticipate many of these problems and utilize shock-alleviating techniques in his designs. Since many shock problems are not predictable, retrofit practices often become a necessity to correct severe shock conditions. It is not uncommon, and the designer should not hesitate to utilize as many of these practices as he finds practical, since over-correction is hardly possible.

PILOT CHOKES

Since sudden blockages or reversals of flow paths in directional control valves is one of the obvious generation points of hydraulic shock, attention should be given to the possibilities of controlling or modulating the speed of shift of the flow director. The most common means of achieving this modulation is by the use of the pilot choke. In its most common form, the pilot choke is a sandwich-type combination of two noncompensated flow-control valves incorporated in a common block. This sandwich block is then placed between the pilot valve and the main directional control valve it pilots. Fig. 9-1 depicts a simple valve and cylinder circuit. The composite valve symbol indicates the relative location of the pilot choke block. As can be seen in tracing the flow pattern, the control is in the pilot lines that supply the directional pilot chambers. Thus, the speed of shifting of the main control valve flow director is controlled by the setting of the flow adjustments within the pilot choke block. These can also be adjusted or regulated for the purpose of effecting a dwell action at one or both ends of the cylinder piston stroke. If the control valve is selected as having a momentary open-center condition at the crossover point, this will further dampen whatever shock tendencies might be inherent

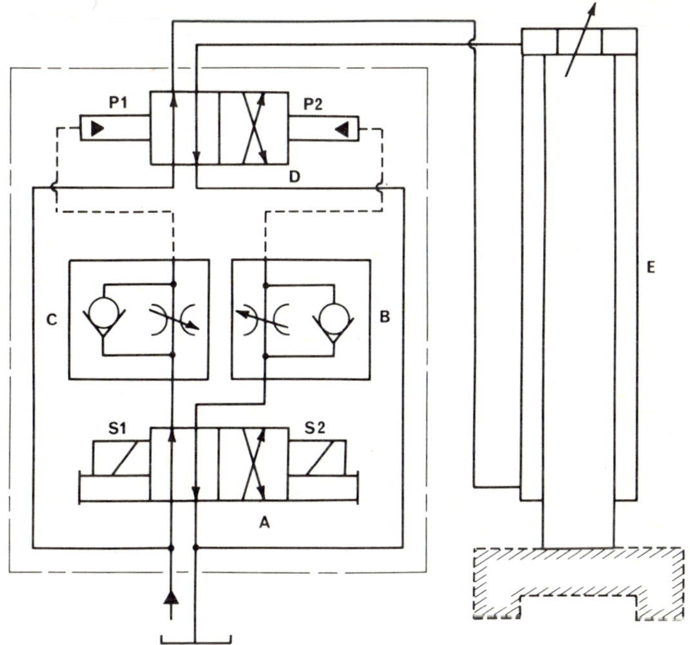

Fig. 9-1. Simple valve and cylinder pilot choke circuit.

in the valve shift. This composite symbol indicates further that this control valve is internally piloted and internally drained to the tank line of the main control valve.

Fig. 9-2 is a circuit similar to that shown in Fig. 9-1, except that one other control variation has been added. The four-way, directional control valve in this illustration incorporates an external pilot supply as well as an external drain connection. The pilot supply to the pilot section is at a reduced pressure, with its supply pressure controlled by pressure-reducing valve R ahead of the pilot supply port. This offers the additional advantage of a reduced shifting pressure for the main piloted flow director. Experience shows that this combination produces excellent results in terms of modulating the shock generation characteristics of a directional control valve.

Another technique often employed for controlling the abruptness of the shear action of a four-way, hydraulic valve flow director is to incorporate beveled or tapered lands on that portion of the valve spool that is involved at the crossover point.

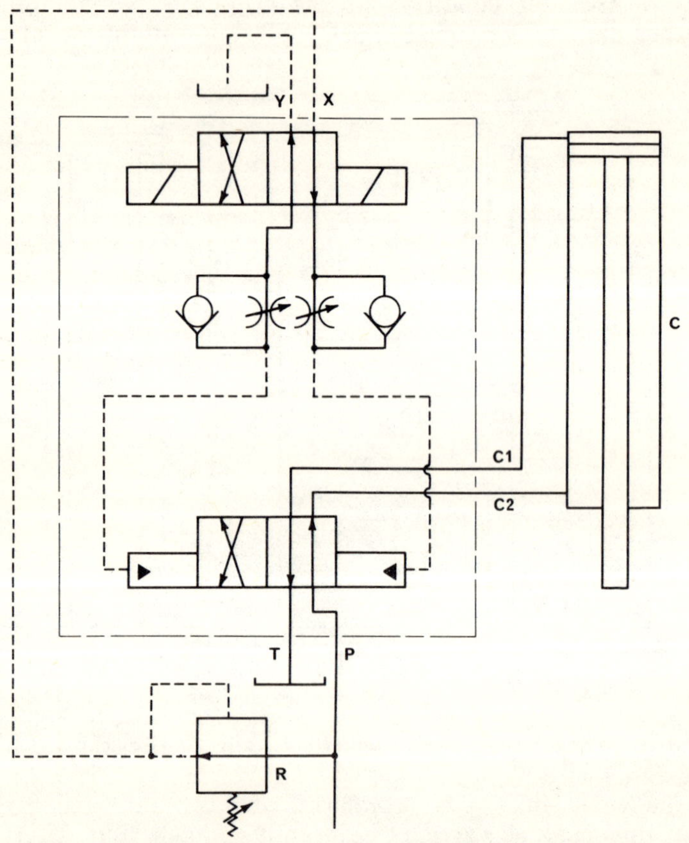

Fig. 9-2. More complex valve and cylinder pilot choke circuit.

While this can be done as an alteration in the field, extreme caution should be employed, as the functioning of the valve could be impaired. It would be wise to check with the supplier of the valve before making such alterations. A variety of techniques have been successfully employed, including the aforementioned beveling of the land edges of spools. A variation of beveling is to create tapered grooves along these edges to give a modulated "pinching off" of the flow path.

An additional gain in shock-modulating capacity can be obtained by the use of flexible hose for some or all of the conduit lines in a hydraulic system. The limited ability of hydraulic hoses to expand with a rise in internal fluid pressure will tend to modulate somewhat the intensity of these transient pressure shocks. While the use of hydraulic hoses will reduce the heat dissipation capacity of a system as compared with the use of steel tubing or other metal conduit lines, the transient shock problem might prove to be of greater significance in a given system.

ACCUMULATORS

With the exception of those systems that employ hydraulic hoses, conduit lines are quite unyielding. If the slight expansibility of hydraulic hose can effect a little shock reduction, greater expansibility will provide for greater shock reduction. The simplest way to provide greater conduit line expansibility is to incorporate an expansion unit into those lines. The hy-

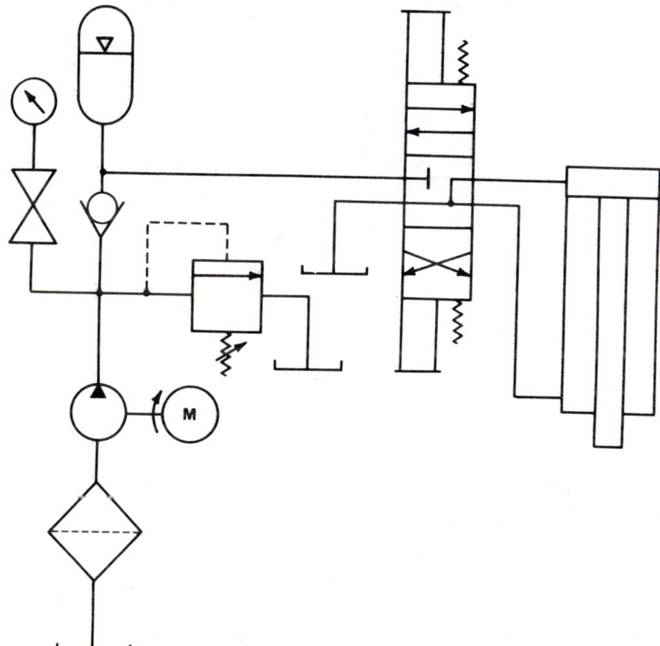

Fig. 9-3. Accumulator used to reduce shock in hydraulic circuit.

draulic accumulator is tailor-made to perform such a function.

The normal functioning of the accumulator involves the movement of liquid into and out of its liquid section. Accompanying this movement is a contraction and expansion of the gas that is captive in its gas section. The compressibility of this gas thus provides a shock-absorbing capacity with a very brief response time. When the accumulator is installed for this purpose, its precharge gas pressure should be approximately ten percent lower than the system operating pressure to ensure activation in anticipation of reactive shock generation.

Fig. 9-3 depicts a simple hydraulic circuit employing a single accumulator installed specifically to perform a shock alleviating function. If a considerable distance exists between the power unit source and the point of usage, additional accumulators should prove to be effective.

CHAPTER 10

Hydraulic Motor Circuits

Hydraulic motors offer a range of torque, speed, and mounting flexibility that is difficult to achieve by electric motors alone. The chief advantages are in the wide range of speeds that can be maintained, the compactness of size in relation to torque, and the ability not only to infinitely vary the speed, but to achieve a stall condition without damage to the componentry.

UNIDIRECTIONAL MOTOR CIRCUITS

Fig. 10-1 depicts a simple, single-direction hydraulic motor circuit with a combination start-stop control, speed control, and torque control. Hydraulic motor MF in this circuit is a fixed-displacement, single-direction motor. Hydraulic fluid supplied by hydraulic pump P is delivered through valve A to the inlet port of hydraulic motor MF. The maximum speed of motor shaft rotation is determined by the displacement of pump P and the displacement of hydraulic motor MF. Since these two displacements are fixed, maximum speed will remain constant, depending upon the load resistance at the hydraulic motor output shaft. A speed less than maximum can be controlled by regulation of needle valve F in the tank line of the hydraulic motor. Maximum torque is determined by the maximum pressure available to the input port of the hydraulic motor. Maximum pressure in the system is determined by the setting of relief valve R. When the two-way, normally closed

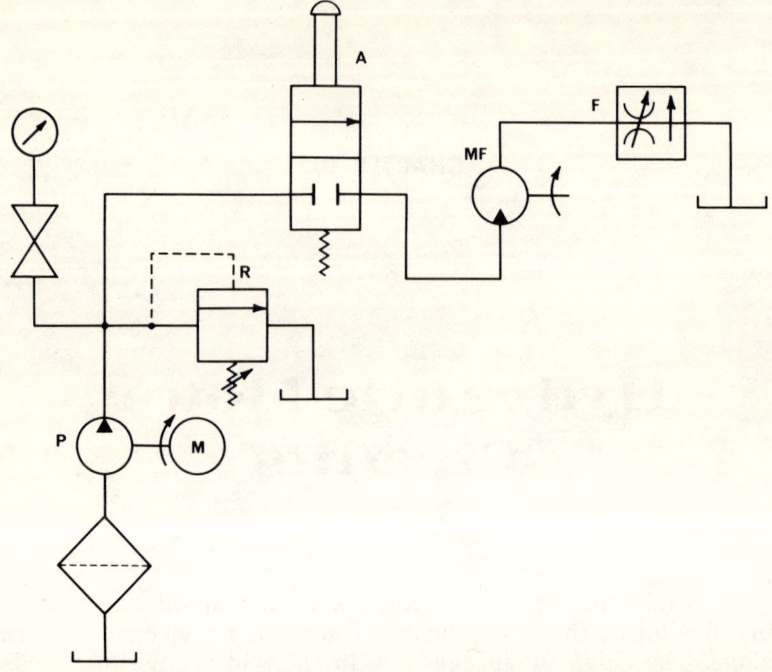

Fig. 10-1. A simple single-direction, hydraulic motor circuit.

valve A is in the spring-offset condition, no fluid is delivered to the hydraulic motor. In this condition, the full volumetric displacement of pump P spills over relief valve R and back to reservoir. The chief advantage of the throttling valve being mounted in the tank line of the hydraulic motor is to control any tendency of the motor to overrun the volumetric delivery rate to its inlet port. With the valve A in the spring-offset position, no fluid is available to the inlet port of the hydraulic

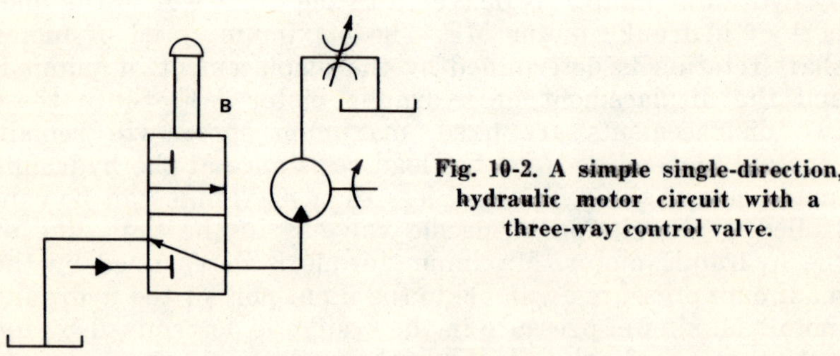

Fig. 10-2. A simple single-direction, hydraulic motor circuit with a three-way control valve.

motor, thus maintaining the hydraulic motor in a semilocked state.

Fig. 10-2 is similar to Fig. 10-1, except that the control valve is a three-way configuration. With control valve B in the position shown, the hydraulic motor is in an unlocked condition and could accept additional rotation of its output shaft.

BIDIRECTIONAL MOTOR CIRCUITS

The abbreviated circuit depicted in Fig. 10-3 incorporates two three-way, normally closed, spring-offset manual valves for supplying fluid to the bidirectional hydraulic motor MF.

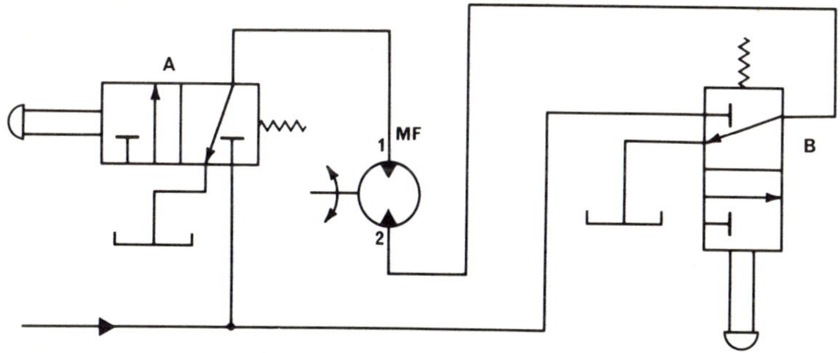

Fig. 10-3. A bidirectional hydraulic motor circuit.

Control valve A, when operated, will supply fluid to the number 1 port of hydraulic motor MF for one direction of rotation. Fluid exhausting from port 2 of motor MF passes through valve B and directly back to the reservoir. When valve B alone is energized, fluid is delivered to the number 2 port of hydraulic motor MF to power the opposite direction of rotation. Fluid exhausting from port number 1 of motor MF returns to the reservoir through valve A. With neither valve A nor valve B operated, the bidirectional motor MF is in a float or unlocked condition and can be rotated by input torque on the output shaft. With both valves A and B operated, the motor is in a pressure-locked condition and will not supply output torque in either direction.

Fig. 10-4 depicts a bidirectional hydraulic motor circuit that is controlled in two directions by the use of a four-way, manually operated, two-position, detented, hydraulic control valve A. The two pressure-compensated, flow-control valves X and Y are used for controlling the speed of hydraulic motor MF in each direction. In this case, the throttling action is performed ahead of the motor and will successfully control the output speed of the hydraulic motor, unless an overrunning load condition is encountered. If such an overrunning load is encountered, the two flow-control valves X and Y should have their direction reversed to accomplish exhaust throttling. The speed capability of the hydraulic motor is equal in each direction. If different speeds are desired in each direction, this

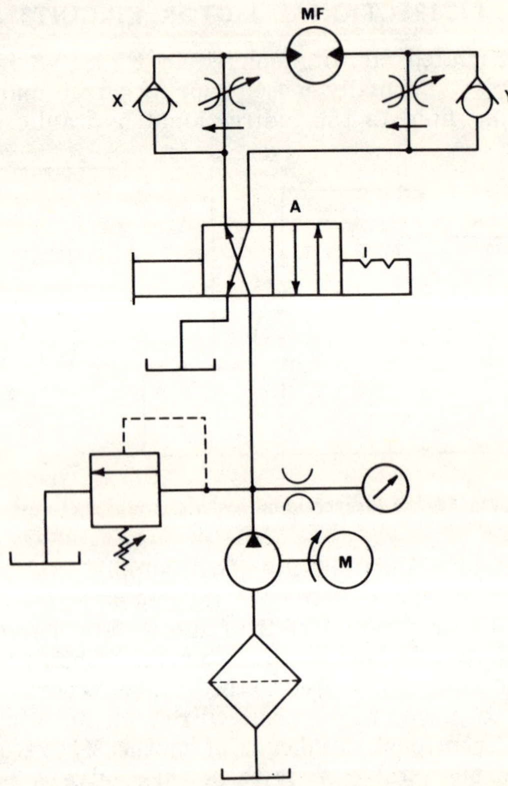

Fig 10-4. A bidirectional hydraulic motor circuit with inlet throttling.

adjustment can be made by regulating the two pressure-compensated, flow-control valves.

Fig. 10-5 offers an additional variation by utilizing a three-position, spring-centered control valve with a center position that connects motor port 1 directly to motor port 2. Bidirectional rotation is achieved by shifting control valve A to one extreme position or the other. When control valve A is in its center position, hydraulic motor MF is in a free-turning, non-pressurized state. If the free-running condition of the motor is

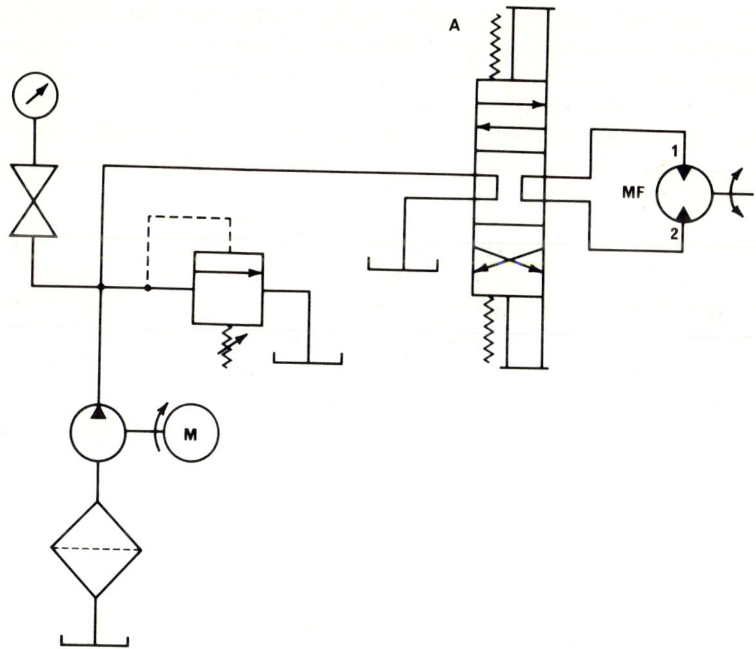

Fig. 10-5. Hydraulic motor circuit with free-running off condition.

not desired, the three-position, tandem-center valve depicted in Fig. 10-6 might be preferable. This tandem-center valve offers a blocked condition for the two hydraulic motor valve ports. If considerable speeds are involved and the inertial load of the hydraulic motor with its attached load is appreciable, attention should be given to Fig. 10-8.

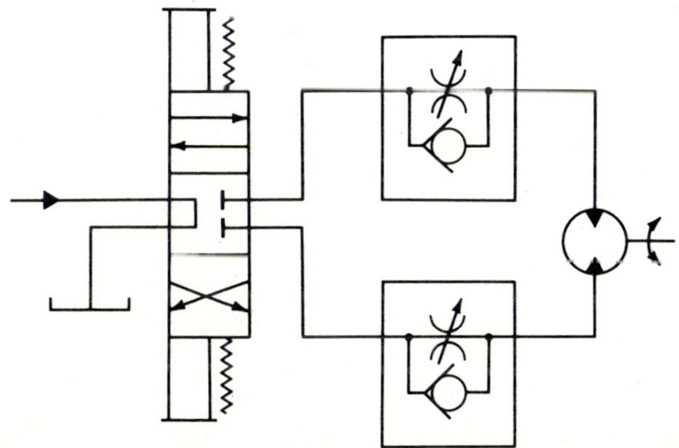

Fig. 10-6. Hydraulic motor circuit with blocked off position.

When two hydraulic motors are being operated from the same power source and the possibility of speed synchronization is desired, Fig. 10-7 offers an interesting and flexible circuit. With the two three-position, tandem-center control valves piped in series and with both valves centered, the pump output is delivered directly through the two tandem centers in series directly back to the reservoir. When either valve is operated by itself, its hydraulic motor will be capable of the full output torque permitted by the maximum pressure value set by relief valve R. With each of the two control valves A and B actuated, the speed of the two hydraulic motors should be nearly identical. With this dual condition, the fluid being delivered to hydraulic motor MF2 has already passed through hydraulic motor MF1. Thus, the speeds of the two hydraulic motors will be matched, but the total torque capacity of a

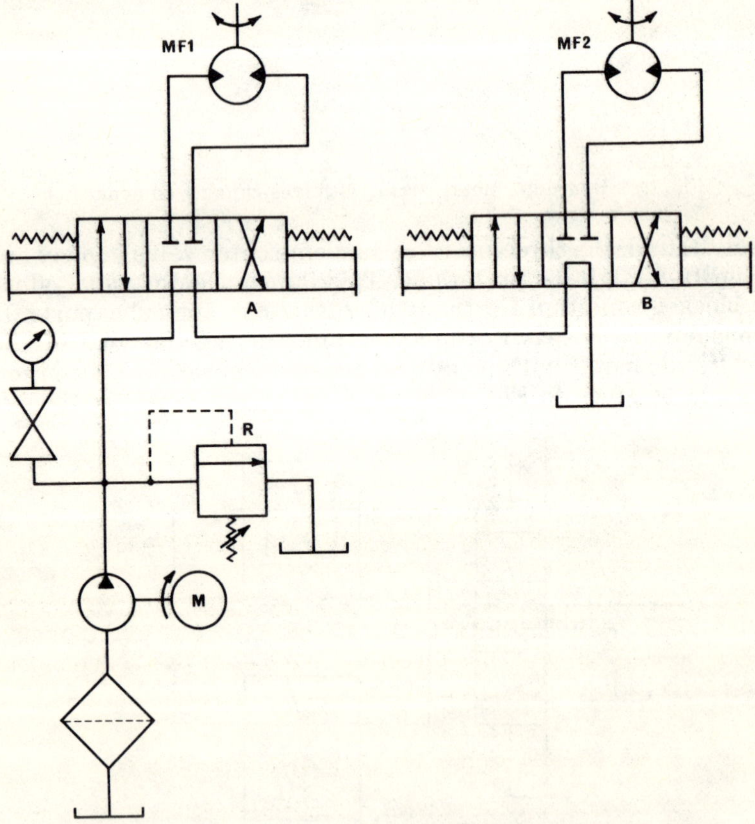

Fig. 10-7. Circuit for operating two bidirectional motors in series.

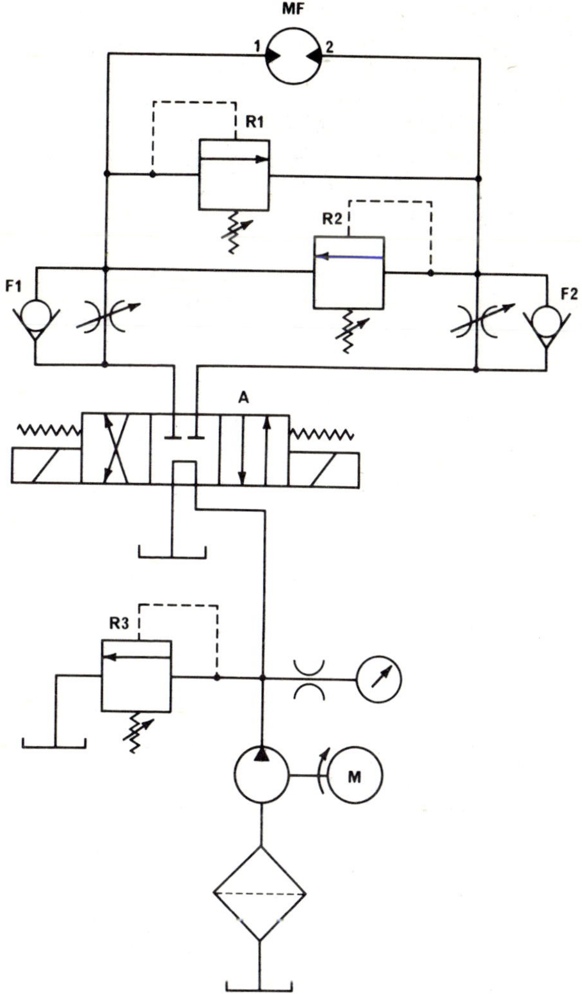

Fig. 10-8. Braking and replenishing with external throttling.

single motor will be divided between the two motors, depending upon the load on the output shafts. These two hydraulic motors may be rotated in either direction, with each one independent of the rotational direction of the other.

BRAKING AND REPLENISHING CIRCUIT

As previously indicated with Fig. 10-6, the necessity of suddenly stopping a bidirectional hydraulic motor may create serious shock loads in the system. Fig. 10-8 depicts a method of controlling the magnitude of these hydraulic shocks. As control valve A is shifted to its center position, both ports of hydraulic motor MF are blocked, bringing the hydraulic motor to a sudden stop. The tendency of the hydraulic motor to continue its rotation, due to the inertia of its attached load creates, a pumping action by the motor with a resultant sudden pressure rise in the downstream conduit lines. The placement of relief valves R1 and R2 tends to limit this pressure buildup. If the pressure in the conduit lines leading from motor port 1 rises, relief valve R1 will open, permitting the fluid to pass directly to line 2. The same is true when relief valve R2 limits the pressure rise in line 2, while suddenly stopping the other direction of rotation. With the two conduit lines blocked at control valve A, relief valves R1 and R2 maintain a constant back-pressure for decellerating the rotation of the hydraulic motor MF. Relief valve R3 controls the maximum pressure that will be generated in the system by the hydraulic pump. Flow-control valves F1 and F2 function as speed controls for hydraulic motor MF and become nonfunctional when control valve A is centered.

HYDROSTATIC TRANSMISSIONS

One of the most versatile means of achieving infinitely variable control over a wide range of output torques and speeds can be found in the hydrostatic-type drive. A full range of bidirectional speeds can be effected with a minimum of components. In its simplest form, a hydrostatic transmission employs an over-center, variable-volume hydraulic pump driving a unidirectional, fixed-volume hydraulic motor, as shown in Fig. 10-9. As the manual control on variable-volume pump PV is manipulated, the pump can be varied from full volumetric displacement from port 1 to full volumetric displacement from port 2. At its midpoint, or crossover point, the pump is delivering no fluid from either port. Manipulation of the manual control can create full speed of hydraulic motor MF in either direction. The full torque capability of the fixed-volume hydraulic motor can be realized at various speeds, as the pressure is controlled by relief valves R1 and R2.

A twin motor drive with an automatic free differential feature is depicted in Fig. 10-10. Two bidirectional hydraulic motors, MF1 and MF2, are supplied by one common over-center, variable-volume hydraulic pump, PV. When rotational adjustment is necessary, as in the cornering of a two-wheel

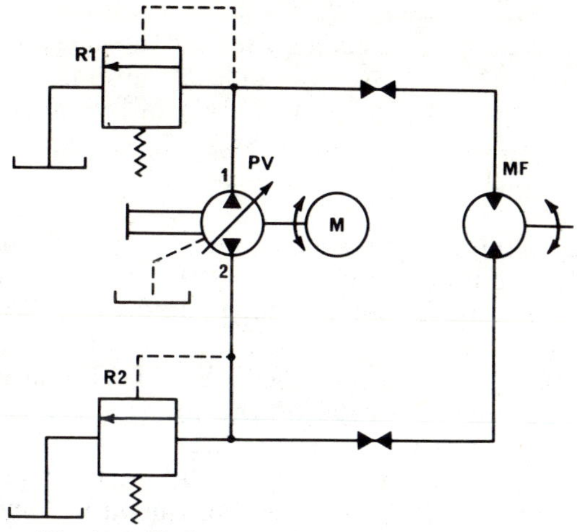

Fig. 10-9. Hydrostatic transmission circuit.

drive vehicle, the proportional flow to the two motors is self-modulating. The torques of the two motors are equal, with the output shaft rotation varying by demand. Speed is determined by the stroke setting of the hydraulic pump, with direction being determined by the direction from center of the pump plate angle. This combination offers a speed variation from zero to full displacement speed in each direction.

A wide range of speed, torque, and control variations are obtainable, depending upon the application requirements. If additional variations are required, attention should be given to other sources that specialize in such transmissions.

VARIABLE-SPEED MOTOR CIRCUITS

Where speed variations in a hydraulic motor circuit are required, attention should be returned to prior illustrations in this chapter where speed-control valves are incorporated. How-

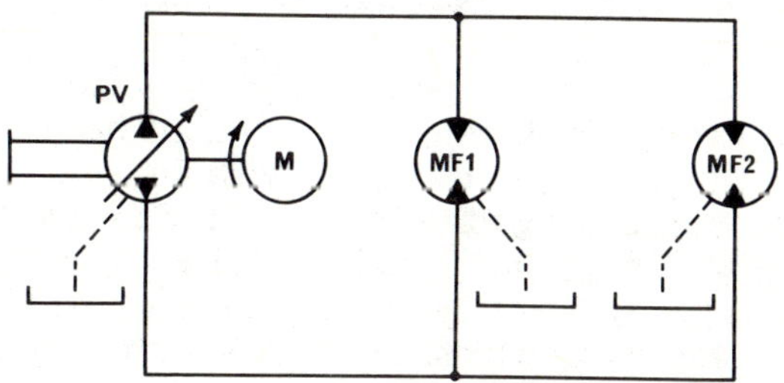

Fig. 10-10. Twin motor drive with automatic free-differential feature.

ever, two additional methods of varying the speed of a hydraulic motor output shaft deserve some attention. Fig. 10-11 resembles, somewhat, the hydrostatic transmission illustration in Fig. 10-9, but there are some noteworthy differences. This illustration depicts a fixed-volume, unidirectional hydraulic pump driving a variable-volume unidirectional hydraulic motor. The maximum pressure value of the fluid being delivered by the pump is determined by the setting of relief valve R. As the volumetric displacement of hydraulic MV is varied, the output speed of the motor shaft will change. Along with the changing of the output speed will be an accompanying change in the output torque of the hydraulic motor. With a given input pressure value, the greater torque will be realized at the lower speed setting. As the volumetric displacement of the hydraulic motor MV is decreased, the output speed will in-

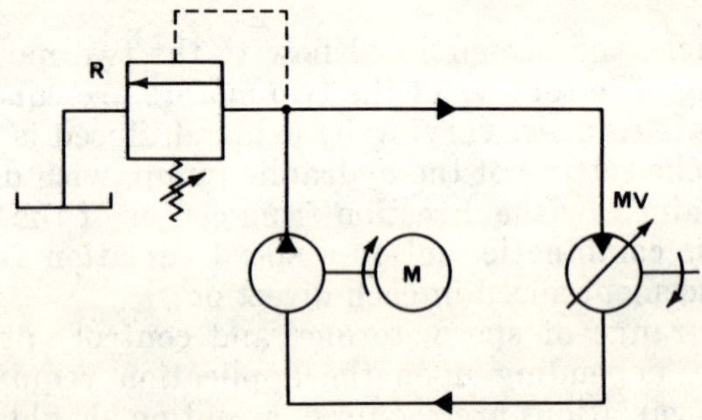

Fig. 10-11. Speed and torque control in a variable-volume, unidirectional hydraulic motor.

crease as the output torque declines. This latter condition is different from the effect realized from the use of flow-control valves that do not alter the volumetric displacement of the motor. The use of this type speed control will be advantageous only where the decreasing torque characteristic is acceptable or desirable.

Fig. 10-12 illustrates bidirectional speed control of a hydraulic motor utilizing the bleed-off technique. The power source is a fixed-volume hydraulic pump with maximum pressure established by the setting of relief valve R. The directional control valve is a three-position valve with the pressure supply blocked in center and both motor ports connected to the tank. Hydraulic motor MF is a fixed-displacement, bidirectional unit. Control of the speed of the motor output is effected by regulating the amount of fluid "bled off" from the input line to

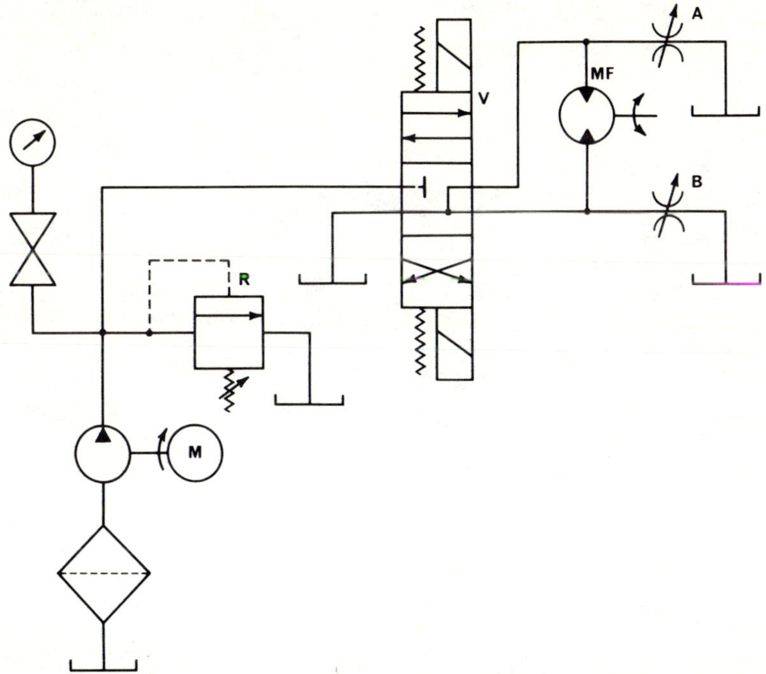

Fig. 10-12. Speed control using bleed-off technique.

each port. While this method affords a wider variation in speed for a given control setting, it may prove desirable for those applications that require a modulated speed reduction as loads increase.

CHAPTER 11

Basic Servo Circuits

The most sophisticated control component available to circuit builders is the servo valve. Unlike the conventional on off-type directional control valves, the servo valve performs the directional function as well as proportional flow control. The traditional on off design approach requires the addition of at least one component for each regulated speed, as well as one additional valve for each change of speed. Acceleration and deceleration usually require even more components. Regardless of the number of components added, true proportional control is not achieved.

CLOSED-LOOP CIRCUITS

The maximum control of speed, direction, and position is readily achievable with a "closed-loop" servo system. This control is maintained by the use of a feedback signal that communicates to the servo valve any discrepancy that develops between the actual performance and the established norm. This feedback line of communication can be mechanical, pressure value, load force, electrical, or a combination of these.

Electrical Feedback

The most versatile and accurate control is obtainable from the electrohydraulic-type unit. Such a unit usually utilizes a dc actuator that is capable of both a positive and a negative signal, as well as a proportional one. The polarity of the signal determines the directional parameter, and the magnitude of the signal determines fluid flow rate.

The feedback signals may originate from a variety of sources, depending on the nature of the parameter being controlled. The sensing element is designed to monitor the parameter and convert its response to a compatible electrical signal. Thus, pressure, speed, or position are each sensed and transduced to an electrical command signal that is compatible with the electrical actuator of the servo control valve. Such a unit is usually referred to as a transducer.

Motor Speed Control

Fig. 11-1 depicts a system in which the rpm of a hydraulic motor is controlled by a servo valve. Motor M could be driving a spindle that requires control of its rotational speed within close tolerances. Tachometer generator T, driven by the spindle,

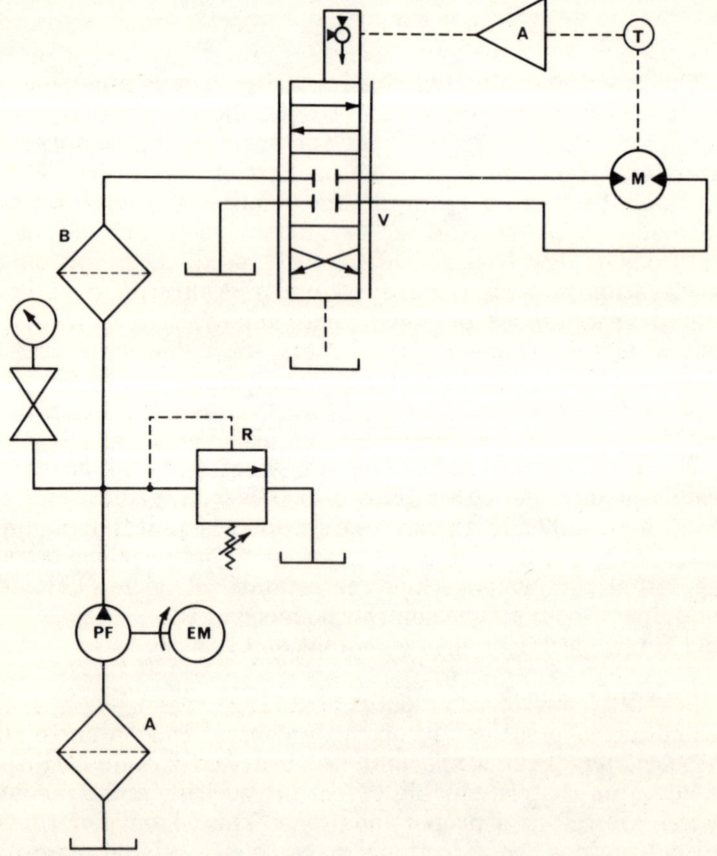

Fig. 11-1. Hydraulic motor speed controlled by a servo valve.

supplies the feedback signal to the force motor of servo valve V through amplifier A. The rpm to be maintained is predetermined by the bias setting of the force motor. When the rpm determination is exactly matched by the tachometer generator output, the bias signal is exactly matched by the feedback signal, and the servo valve is in a stabilized condition. Any variation from the predetermined speed of the spindle will cause the tachometer generator to feed back a signal of a greater or lesser magnitude, with a resulting response by the servo valve force motor. The tachometer generator in this circuit is an rpm transducer.

Servo valve V is capable of infinite positioning. Relief valve R establishes the maximum system pressure, with volume supplied by constant- or fixed-displacement pump PF. Two filters are contained in the circuit, with A being of the suction-strainer type to protect the pump. The small, sensitive orifices found in the servo-type valve require finer filtration than can normally be supplied by a suction-type filter, hence the inclusion of pressure filter B. Pressure filter B must be capable of withstanding full system pressure and should stop all particles down to 10-micron size. Some servo valve manufacturers request as low as 5-micron filtration ahead of the servo unit.

Linear Force Control

Fig. 11-2 depicts a linear force unit (hydraulic cylinder C) controlled by servo valve V. The parameter being controlled is the force exerted by the cylinder C piston. The sensing unit is a force transducer, sometimes called a "load cell." Force transducer FT is capable of generating a proportional signal based on the force load transmitted through it. When the magnitude of its signal matches the predetermined input signal to the servo force motor, the servo valve is stabilized until a signal change occurs. The input-signal device is not shown, but it could assume a variety of configurations, such as push button, selector switch, or even tape control.

OPEN-LOOP CIRCUITS

Presently, automation demands a large variety of hydraulic speed-control circuits. It is posible to use servo valves in open-loop velocity systems, and proportional control is obtained by varying the current to the force motor. But, since the servo valve is pressure sensitive, the flow varies when the load pressure changes. Closing the loop with velocity feedback will eliminate the problem, but the solution is expensive.

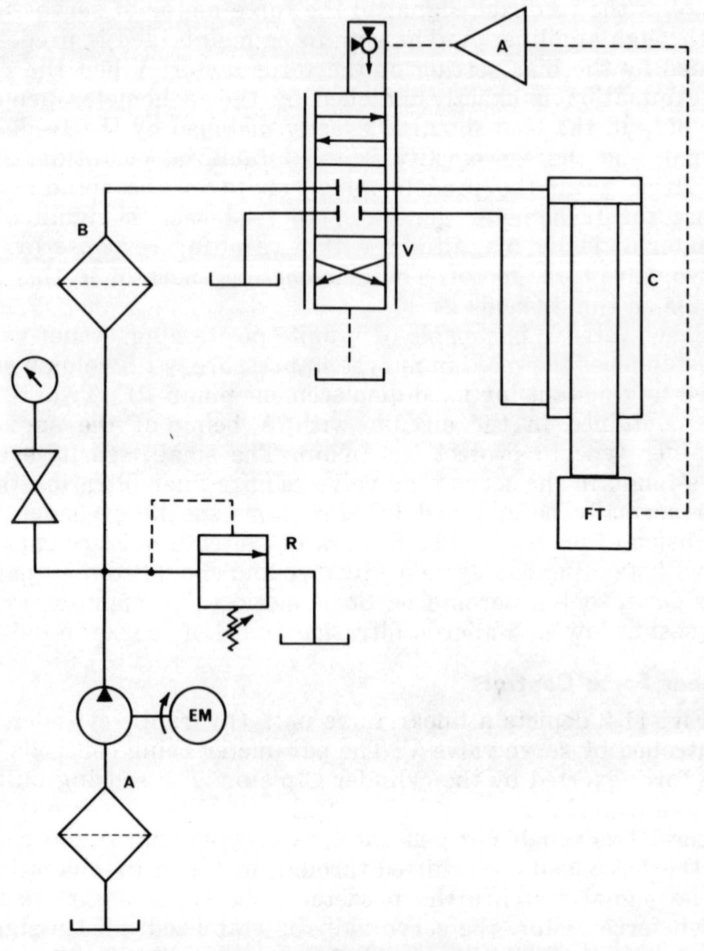

Fig. 11-2. Servo valve control of hydraulic, linear-force unit.

The problem of variable loads, as well as the expense of the feedback linkage, dictates the use of a four-way, electrohydraulic, proportional flow, pressure-insensitive control valve. One valve, properly connected in a circuit, will replace a four-way valve, counterbalance valve, and as many flow controls and deceleration valves as are needed in the circuit.

Many applications require proportional speed control, accurate regulation of speed (flow), and programmable changes of the stabilized performance value. Fig. 11-3 depicts a hydraulic circuit whose performance requires multiple speeds as well as acceleration and deceleration in each direction. Torque-motor operated, four-way valve V performs all the functions shown in Figs. 11-4 and Fig. 11-5. The electronic programmer offers a variety of channels, each with its own signal magnitude determinant. Flow is determined by the current to the dc torque-motor operator and is not affected by pressure variation due to load changes. Any number of speeds, accelerations, or decelerations in both directions can be achieved by merely increasing or decreasing the voltage level or rate of increase or decrease of voltage to the torque motor. A timed program may be employed, or limit switches may be used to select channels if speed changes or decelerations must start at exact points of the cylinder piston travel. In addition to controlling velocity,

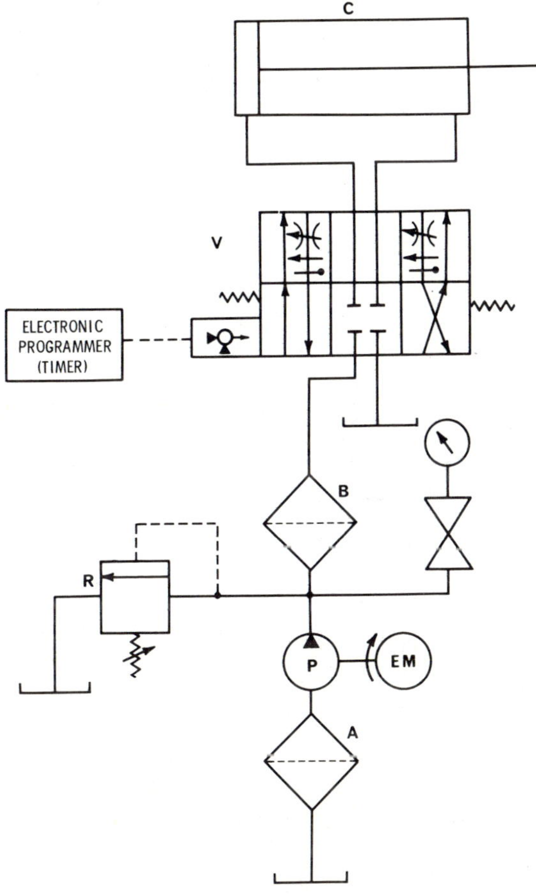

Fig. 11-3. Illustration showing electronic programming of a hydraulic circuit.

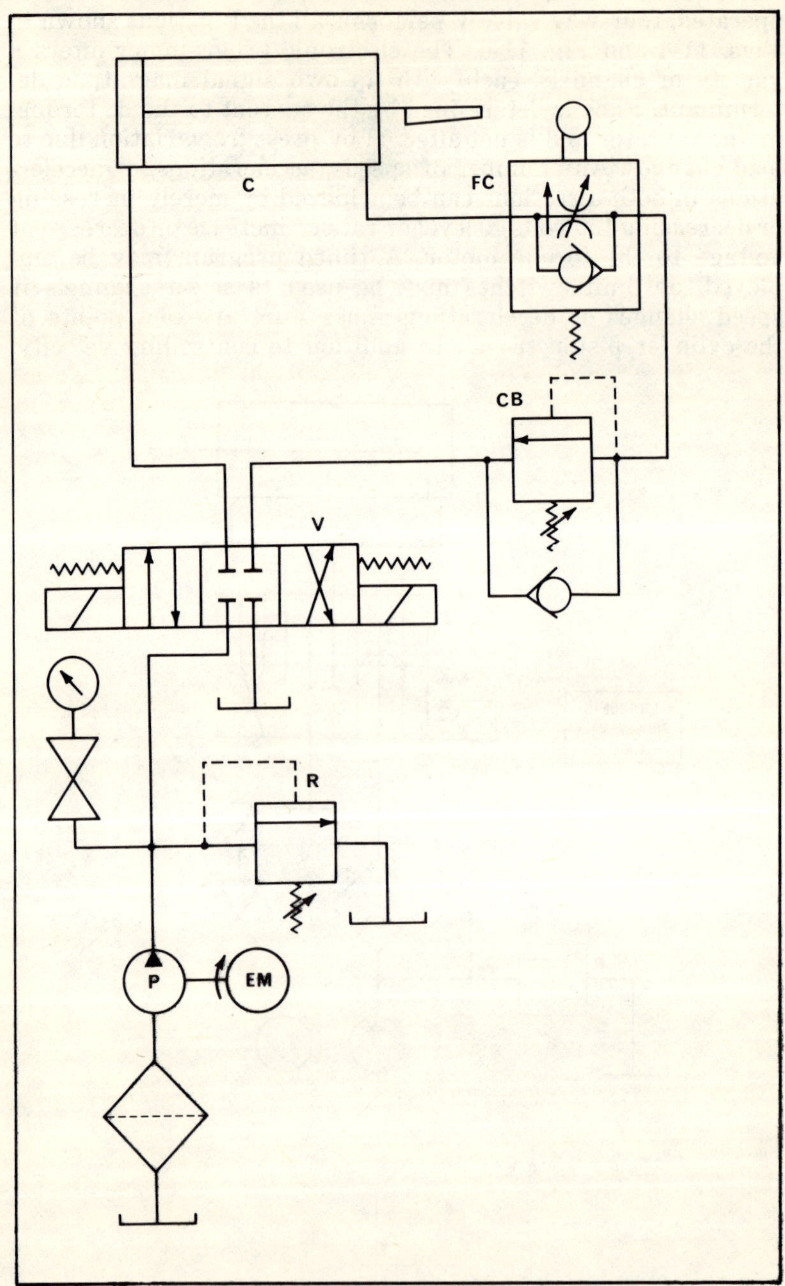

Fig. 11-4. Deceleration circuit.

acceleration, and deceleration, this valve will control overhauling loads as well.

The control features of this system are as follows:

1. Flow is proportional to input current.
2. Acceleration can be controlled.
3. Deceleration can be controlled more accurately.
4. Flow rate can be controlled from a remote location.
5. Flow rate can be changed in milliseconds.

Counterbalancing a load or stopping an overrunning load is part of the capability of such a valve. The valve is both a meter-in and meter-out flow control. When the input pressure is reduced because of an overrunning or overhauling load, the

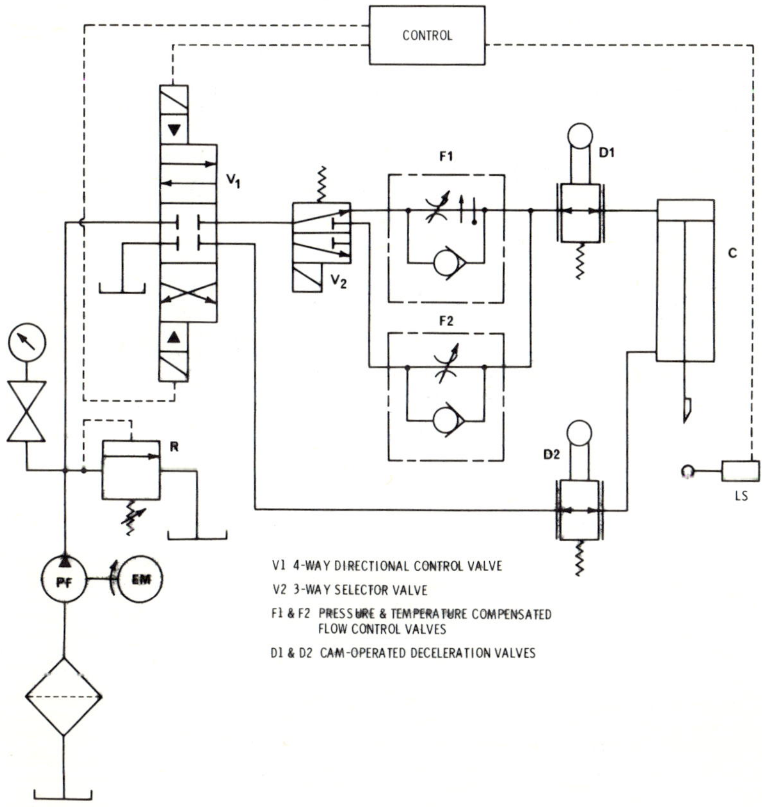

Fig. 11-5. Circuit for multiple speed with deceleration.

output restriction builds up instantly, holding the load speed at its set point. The flow rate is adjustable all the way to zero. There is no question of valve synchronization, since both the amplitude and direction of the flow are controlled by the same spool.

The spindle drive circuit depicted in Fig. 11-1 would assume the configuration shown in Fig. 11-6. The chief difference between the two circuits is the absence of the transducer and feedback linkage depicted in Fig. 11-1. The Fig. 11-6 circuit will offer rotational accuracy sufficient for most spindle drives due to insensitivity of the valve to load changes. The open-loop

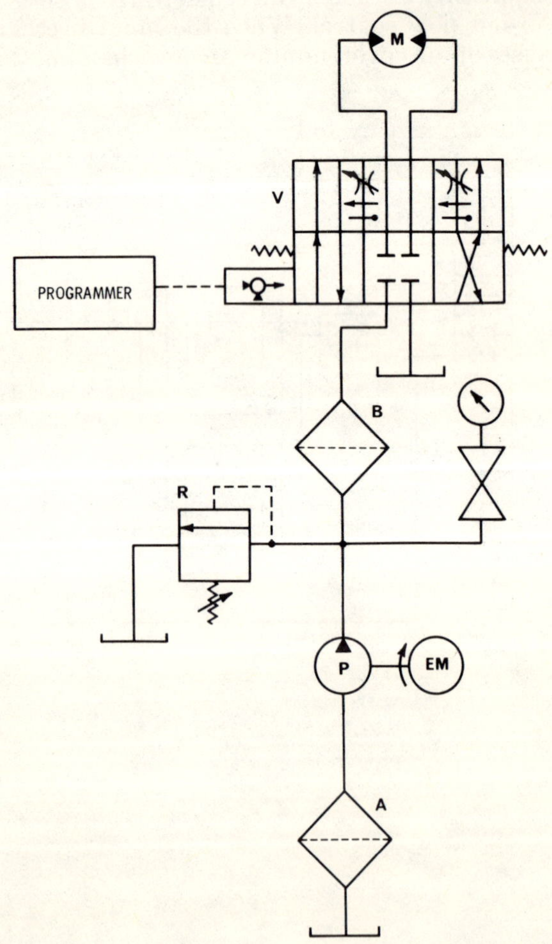

Fig. 11-6. Circuit similar to Fig. 11-1 without feedback linkage.

accuracy of this valve is conservatively rated at ±2 percent of set point for a 50 percent load change.

Each circuit depicted in the preceding chapters represents a concept and is not necessarily the only valving arrangement available to the circuit builder. By maintaining component flexibility and using ASA symbology, the imaginative circuit designer can create and prove the type of component combination that will result in the degree of control the application requires. Good luck!

Index